Progress and Issues in Transgenic Plants

Progress and Issues in Transgenic Plants

Edited by **Harvey Parker**

New York

Published by Callisto Reference,
106 Park Avenue, Suite 200,
New York, NY 10016, USA
www.callistoreference.com

Progress and Issues in Transgenic Plants
Edited by Harvey Parker

International Standard Book Number: 978-1-63239-518-4 (Hardback)

Printed in the United States of America.

Contents

Preface

Every book is initially just a concept; it takes months of research and hard work to give it the final shape in which the readers receive it. In its early stages, this book also went through rigorous reviewing. The notable contributions made by experts from across the globe were first molded into patterned chapters and then arranged in a sensibly sequential manner to bring out the best results.

This book on transgenic plants is a compilation of information by veteran researchers from across the world. Progress of effective transformation protocols is becoming an integral strategy to traditional breeding methods for the enhancement of crops. This book evaluates the effect of genetically transformed crops on biosafety. It is structured into two sections namely, Metabolomics and Biosafety. It would be beneficial to students, researchers and botanists.

It has been my immense pleasure to be a part of this project and to contribute my years of learning in such a meaningful form. I would like to take this opportunity to thank all the people who have been associated with the completion of this book at any step.

Editor

Part 1

Metabolomics

Transgenic Plants as Gene-Discovery Tools

Yingying Meng[1], Hongyu Li[1], Tao Zhao[1], Chunyu Zhang[1],
Chentao Lin[2] and Bin Liu[1,*]
[1]Institute of Crop Science, Chinese Academy of
Agricultural Sciences, Beijing
[2]Department of Molecular, Cell & Developmental Biology
University of California, Los Angeles,
[1]China
[2]USA

1. Introduction

Mutation study is an important strategy to dissect gene functions. Classical chemical or irradiation mutagenesis is one of the most powerful screen approaches to uncover genes involved in certain genetic pathways. However, the tedious works to hunt down the gene corresponding to a mutant allele by map-base cloning make an intrinsic limitation of this approach. Loss-of-function mutations resulting from Transferred DNA (T-DNA) insertion or Mobile Genetic Elements (MGE) insertion have overcome this shortcoming. Transgenic approaches are also useful to circumvent the difficulties in the study of gene function because of genetic redundancy or lethality. Gain-of-function mutations achieved by activated expression of endogenous genes by transcription enhancer or by specific gene over-expression through transformation have also revealed function of many plant genes. This review will describe the major loss or gain-of-function mutagenesis approaches by T-DNA vector transformation or endogenous MGE and those available insertion mutant resources which have greatly facilitated the researchers to extensively identify gene functions in the past few years.

2. Loss-of-function mutations

The most conventional tool for the functional analysis of all genes in a certain organism is to generate indexed loss-of-function mutagenesis on a whole genome scale. For example, the creation of gene-indexed loss-of-function mutations for all genes has been achieved decades ago in the unicellular budding yeast *Saccharomyces cerevisiae* by gene replacement via homologous recombination (Ross-Macdonald et al. 1999; Winzeler et al. 1999; Giaever et al. 2002). However, genome wide gene disruptions with confirmed index information are not so easy for the multi-cellular eukaryotes. Loss-of-function can be achieved by chemicals

---

*Corresponding Author

such as ethyl methannesulfonate (EMS) or by high energy irradiation such as fast neutrons to introduce random mutations (Ostergaard and Yanofsky 2004). These methods are easy to generate large scaled mutagenesis, but the mutation site identifications by traditional forward genetic method need tedious work. Although popular to individual research groups who focus on certain aspect of their interested field, such kinds of methods are not convenient for systematic research of gene functions on a genome wide scale.

Loss-of-function mutations achieved by T-DNA insertion or MGE insertion provide practicable methods to identify the genes disrupted by these transfer elements because their insertion sites can be explored by PCR base methods such as tail-PCR, plasmid rescue, inverse PCR or adapter PCR (Liu et al. 1995; Liu and Whittier 1995; Spertini et al. 1999; Yamamoto et al. 2003). *Arabidopsis* is currently the only multi-cellular organism reported possible to achieve saturation of mutation with accurate index for each genes because *Arabidopsis* owns several crucial advantages than other species: genome with small size, easily to be transformed, self-pollination, short life cycle and bulk storage of seeds. Rice is the second better learned species that has been widely used for establishing systematic insertion mutant libraries because rice is one of the most important crops and easy to be transformed by *Agrobacterium* mediated transformation. So this review will introduce the basic strategies of the major transgenic approaches using *Arabidopsis* and rice as model plants and summary the emerged transgenic plant resources which have become the most powerful gene discovery tools in the plant kingdom.

2.1 Saturated T-DNA insertion

The ultimate goal of genome research is to characterize the function of all the genes. As the major step towards this goal, the genome sequencing projects of several plant species were already completed. However, functional genomics studies are still in progress to deduce the functions of all the sequenced genes. Saturation mutagenesis by T-DNA insertion is one of the most successful approaches for analysis of systematic gene functions in the past decade years benefitting from the continuous improvement of transgenic techniques (Parinov and Sundaresan 2000). The *Agrobacterium* vacuum infiltration method for *Arabidopsis* transformation was developed to avoid the complex transgenic process of tissue culture based method including introduction of DNA by particle bombardment or *Agrobacterium* and plant regeneration (Bechtold 2003). This method was further modified and the transgenic *Arabidopsis* can be easy achieved by simply dip the floral tissue into a solution containing sucrose, surfactant Silwet L-77 and *Agrobacterium tumefaciens* carrying target genes (Clough and Bent 1998). The basic scheme for the generation of T-DNA insertion mutants by flora dip method for *Arabidopsis* is shown in Figure 1. *Agrobacterium* harbouring the binary vector was used to transform *Arabidopsis* by floral dip method. The T-DNA region between left border (LB) and right border (RB) of the binary vector was randomly transformed into the genome of host plant. The transgenic plants can be screened by selection marker depends on the T-DNA vector used (For example: *NPTII* is the selection marker of pROK2 T-DNA vector which is used to generate Salk insertion mutant lines). The flanking sequence at both sides of insertion can be further sequenced by PCR based method.Taken the advantage of flora dip method, a vast number of T-DNA insertion lines have been generated which represent almost saturated insertions into *Arabidopsis* genes in past ten years.

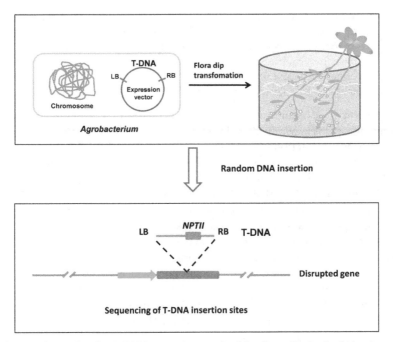

Fig. 1. A basic scheme for the T-DNA insertion method by flora dip in *Arabidopsis*.

2.2 The enhancer trap system

Besides to disrupt the gene functions by direct insertion for large-scale discovery of gene function, T-DNA can also be utilized to identify novel regulatory elements. As a further development of T-DNA random insertion strategy, the enhancer trap system is established by random integration of a report gene cassette as the T-DNA into the genome. This reporter cassette includes a minimum or truncated promoter which is not able to drive the expression of the report gene unless activated by the endogenous regulatory sequence close to the integration site (Figure 2). Therefore, the expression of reporter gene implies the existence of enhancer element close to the insertion site. Because the T-DNA cassette of enhancer trap system can both generate gene mutations and report the presence of enhance element around the insertion site, the enhancer trap system was widely applied in the bacterium, Drosophila, *Arabidopsis*, moss and rice to unveil gene functions and identify regulator elements (Casadaban and Cohen 1979; Sundaresan et al. 1995; Bellen 1999; Campisi et al. 1999; Hiwatashi et al. 2001; Zhang et al. 2006).

As a successful example, the Rice Mutant Database (RMD) is established with the enhancer trap system and maintained by National Center of Plant Gene Research (Wuhan) at Huazhong Agricultural University. The enhancer trap T-DNA fragment used by RMD carries three critical components as indicated in Figure 2:

a. -48 CaMV minimum promoter. This truncated promoter is unable to drive gene expression unless there is a transcription enhancer element close to it.

b. A recombinant DNA sequence encoding an artificial transcriptional activator by the combining of GAL4 and VP16. GAL4 is a DNA bind domain which specifically binds to the Upstream Activator Sequence (UAS) and V16 is a transcriptional activator domain derived from a herpesvirus protein (Utley et al. 1998) which is able to activate the expression of gene adjacent to the UAS.

c. The *β-Glucuronidase* (*GUS*) report gene downstream of the 6 tandems UAS (6xUAS).

The enhancer trap T-DNA including these three components is randomly integrated into host genome by *Agrobacterium*-mediated transformation. If the integration site is by chance neighbour to a host transcriptional enhancer element, the -48 CaMV minimum promoter will drive the expression of the GAL4/VP16 recombinant transcriptional activator which will then bind to 6xUAS and activate the expression of the GUS report gene. Some of the enhancer elements regulate spatial- or temporal gene expression and thus the transgenic lines with the T-DNA inserted near these enhancer element show expression of the report gene in a tissue specific pattern. Besides causing gene mutations and providing an efficient approach for identification of the transcriptional enhancer, enhancer trap T-DNA can create certain pattern lines which are useful to ectopic express target genes in certain tissues simply by cross the pattern lines with the target lines which are transformed by the target genes driven by 6xUAS. Figure 2 shows the basic scheme for the enhance trap system: The T-DNA of enhance trap vector includes -48 CaMV promoter, *GAL4/VP64* transcription activator and *GUS* reporter under control of 6xUAS. The transgenic lines were obtained by *Agrobacterium* mediated transformation of rice callus. The enhance trap T-DNA was randomly integrated into the rice genome and the enhancer nearby the insertion site activates the expression of *GAL4/VP64* which further promote the expression of *GUS* report gene in the Pattern line. The Target line was created through the transformation of host plant with T-DNA containing 6xUAS :: *Target gene*. By crossing Target line with Pattern line, the target gene will express in the same pattern as that of the *GUS* report gene due to the trans-activation of 6xUAS regulator by GAL4/VP64 transcription activator.

2.3 Mobile Genetic Elements (MGE) insertion

In addition to T-DNA insertion lines, MGE insertion is also a popular approach to generate large number of mutations. MGEs which can move around within the genome include several kinds of mobile DNA elements such as transposon or retrotransposon. Transposons describe the DNA which can be cut away from one site and paste to other place within the genome. Retrotransposon however, make themselves a copy and then paste to other position within the genome.

Several transposable elements identified in maize have been used to obtain large population of insertions in genes for functional genomics studies. For example, the maize transposable element *Activator* (*Ac*) first identified by McClintock (Mc 1950) is a kind of transposon widely used for creating MGE insertions. *Ac* element can insert themselves into genes and cause insertion mutations to create a recessive allele. The mutations caused this way are unstable because the *Ac* element can be excised from the inserted gene by the transposase which is coded by Ac element itself. *Dissociation* (*Ds*) element is usually stable because they are incapable of excising itself from the inserted gene unless with the help of *Ac* element.

Researchers combine these two mobile elements and named it as the *Ac/Ds* system to generate mutant populations. Generally, the individual *Ds* parental lines and *Ac* parental lines are created by transformation of *Ac* element and *Ds* element independently into the host organism. Then the two parental lines are crossed to induce the translocations of the *Ds* element in the next generation. For example as shown in Figure 3, *Ac* parent line and *Ds* parent line are created by transformation of host plants with the *Ac* element and *Ds* element respectively. By crossing *Ac* parent line and *Ds* parent line, *Ds* element is activated by *Ds* element to transfer from one position to another position within the genome which will create random disruption of gene functions in the following generations. Stable *Ds* insertion mutant lines can be created by genetic method to make *Ac* element segregated away with *Ds* element by combination of a positive selection marker on *Ds* element and a negative selection marker on *Ac* element (Sundaresan et al. 1995). Besides in maize, the *Ac* element has shown translocation activity in *Arabidopsis* (Fedoroff and Smith 1993). With the *AC/Ds* or other similar MGE systems, several research groups have generated mutant resources with a high proportion of single-copy transposon insertions (Sundaresan et al. 1995; Martienssen 1998; Tissier et al. 1999; Ito et al. 2002; Kuromori et al. 2004; Ito et al. 2005; Nishal et al. 2005)

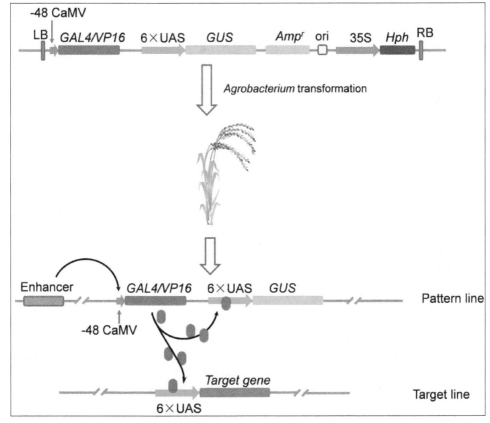

Fig. 2. Overview for the enhance trap system used by RMD.

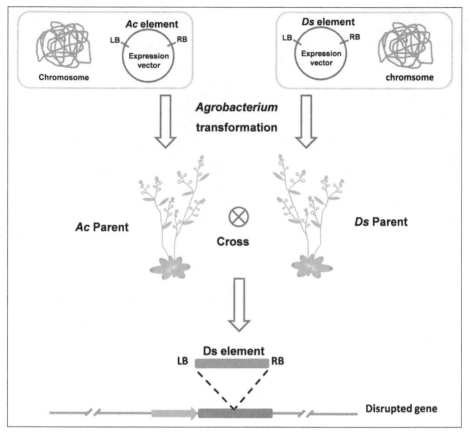

Fig. 3. Scheme for the generation of insertion mutant lines by *Ds* transposon.

Tos17 is one kind of copia-like retrotransposons in rice (Hirochika et al. 1996) ,which can duplicate and paste to elsewhere in the genome. *Tos17* owns several special features that make it suitable for engineering large scale insertion mutagenesis:

a. The copy number of *tos17* is quiet low, ranging from one to five among rice cultivars. For example, the genome of cv. Nipponbare, the selected cultivar for the IRGSP (International Rice Genome Sequencing Project)(Sasaki and Burr 2000), contains only two native copies of *tos17*.

b. Transposition of tos17 is inactive under normal conditions but only activated in the callus by tissue culture and then becoming stable again in the regenerated plants (Hirochika et al. 1996; Miyao et al. 2003).

c. The transposition site of *tos17* prefers gene-dense regions over centromeric heterochromatin regions with a three times higher insertion frequency in genic regions than in intergenic regions (Hirochika et al. 1996; Hirochika 2001; Miyao et al. 2003; Piffanelli et al. 2007).

d. Its size is just a little bit over 4kb and its insertion sequence is clearly known for flanking sequencing.

Tos17 is stably present in genome during the normal life cycle of rice. By the tissue culture of the rice callus, the transcription of *tos17* is activated and the reverse transcript DNA fragments are integrated into new places in the genome which creates disruption of genes. The original *tos17* and its duplications become silence again in the regenerated insertion mutant plants (Figure 4). Taking these advantages of *tos17*, Large-scale *tos17* T-DNA mutant library of Nipponbare has been established by tissue culture and stably preserved by normal generation (Miyao et al. 2003; Sallaud et al. 2004; Miyao et al. 2007; Piffanelli et al. 2007).

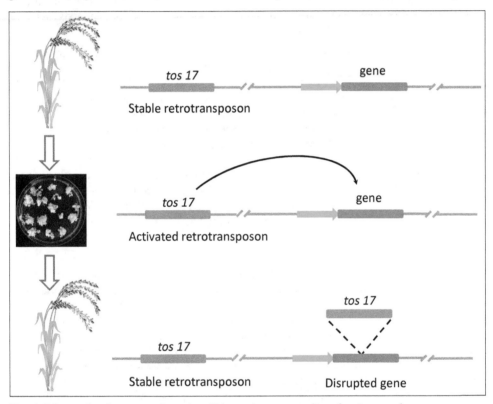

Fig. 4. Scheme for the generation of *tos17* insertion mutant lines by tissue culture.

3. Gain-of-function

Screen for loss-off-function mutations is a primary tool for dissecting a genetic pathway. However, because many genes belong to gene families, loss-of-function screens are not always possible to identify genes that act redundantly. In addition, some genes are critically required for the survival of plants. The homozygote mutants of these genes will not be available for entire function research because of embryonic or gametophytic lethality. As another option, gain-of-function technologies were developed to compensate the limitations of loss-of-function approaches or confer new function in transgenic plants, which is achieved through activation expression of endogenous genes by transcription enhancer which is randomly introduced into the genome or through ectopic gene over-expression driven by constitutive promoter.

3.1 Activation tagging

Activation tagging is a gain-of-function method that generates transgenic plants by T-DNA vectors with tetrameric cauliflower mosaic virus (CaMV) 35S enhancers which can lead to an enhancement expression of adjacent genes in the distance ranging between 0.4 to 3.6kb from the insertion site (Weigel et al. 2000). Differently from the action of the complete CaMV 35S promoter, CaMV 35S enhancers can activate both the upstream and downstream gene transcription. In addition, it has been reported that in at least one case, rather than led to constitutive ectopic expression, CaMV 35S enhancers elevate transcriptional activity based on the native gene expression pattern (Weigel et al. 2000).

Activation tagging technique was firstly developed by Walden and colleagues in decades years ago (Hayashi et al., 1992). Since then, several large scale activation tagging mutant resources have been generated and activation tagging method was widely used to isolate new genes. As an early example, the activation-tagging technique was used in tissue culture to identify cytokinin-independent mutants in *Arabidopsis* and *CKI1* gene whose overexpression can bypass the requirement for cytokinin in the regeneration of shoots was identified (Kakimoto, 1996). Based on the original activation tagging vectors, Weigel and colleagues (2000) developed new generation of vectors possessing resistance to the antibiotic kanamycin or herbicide glufosinate which is low toxic to humans or easy to select transgenic plants in soil in large scale. By screening a set of the transgenic lines, they identified 11 dominant mutants with obviously morphological phenotypes and 9 of them were confirmed due to the activation of adjunct genes by reproducing the phenotype in a new set of transgenic lines through overexpression of the adjunct candidate genes on both sides of insertion.

To accelerate the recapitulation process of phenotype resulted from the enhancer of T-DNA insertion, a new activation-tagging method has been developed using a pair of plasmids including pEnLOX and pCre. pEnLOX contains multimerized CaMV 35S transcriptional enhancers flanked by two *lox P* sites on both sides while pCre includes the *cre* gene which can remove the DNA sequence between two *lox P* sites (Pogorelko et al. 2008). the activation-tagging lines containing the pEnLOX were named the E-lines, and the helper lines containing pCre was named the C-lines. By crossing the E-lines with the C-lines, the CaMV 35S enhancers can be removed from the chromosome coming from E-lines and thus the reversion from mutant phenotypes to the wild-type phenotype may be detected in the next generation.

Activation tagging has also been applied to generate rice activation-tagging lines (Jeong et al. 2002; An et al. 2005; Jeong et al. 2006; Hsing et al. 2007; Wan et al. 2009). Based on the basic activation tagging technology, a dual function T-DNA vectors have been developed for both promoter trapping and CaMV 35S enhancers activation tagging(Jeong et al. 2002). By analysis of the gene expression in these rice activation tagging lines, the authors reported the activated the expression of genes located up to 10.7 kb from insertion site of the enhancers (Jeong et al. 2006). The activation tagging vector pSK1015 is used as an example in Figure 5. The T-DNA contains *BAR* gene as transgenic plant selection marker and 4x35 enhancers as activation element. The activation tagging T-DNA is integrated into the genome of host plants by *Agrobacterium* mediated transformation. The expressions of both side genes around the inserted T-DNA enhancer are elevated.

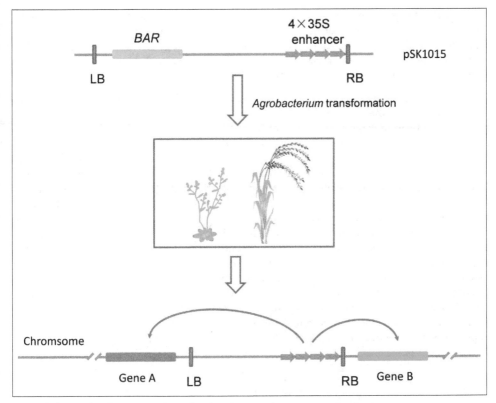

Fig. 5. Scheme for the generation of activation tagging transgenic plants.

3.2 Fox hunting system

FOX hunting system (full-length cDNA overexpressor gene hunting system) is used to ectopic expression of full-length cDNAs (fl-cDNA) in plants to generate systematic gain-of-function mutant populations (Figure 6). The most significant difference between FOX hunting system and other transgenic gene discovery methods is that FOX need to construct a large number of expression vectors containing as many as possible the independent fl-cDNAs, while other methods generally depend on one or a small number of expression vectors. Although the construction of numerous vectors seems labour-intensive, FOX hunting system shows unique benefits in discovery of new gene functions and potential utility of heterologous genes in improvement of agronomic traits.

Firstly, FOX hunting system can be flexibly used to systematically research of gene functions by generation of transgenic plants (also named FOX lines) expressed the fl-cDNA derived from the same species or heterologous host.

The first systematic gain-of-function transgenic population by FOX hunting system was produced by overexpression of *Arabidopsis* fl-cDNAs in *Arabidopsis* and these FOX lines showed various physiological and morphological phenotypes. Thus far, FOX approach has been applied to generate several other FOX population using *Arabidopsis* or rice as model plants by the transformation of rice with rice fl-cDNA or the transformation of *Arabidopsis* with rice fl-cDNA. In addition, FOX hunting technology is an idea approach to systematically investigate the gene functions for those plants such as maize and wheat, the genome of which are too large or not easy to be sequenced.

Secondly, FOX hunting system is powerful to identify lead genes from relative or distant species to improve the traits of plants. It's interesting that the *Arabidopsis* FOX lines with ectopic expression of some rice genes that has no homolog gene in *Arabidopsis* also display abnormal phenotypes, which demonstrates that heterologous gene expression could introduce new function among different species. FOX approach has also been used to screen salt stress tolerance genes in salt cress (*Thellungiella halophila*). As a well know example of Monsanto's huge success by the similar approach, the gene (*BAR* or *PAT*) resistance to herbicide, firstly isolated from *Streptomyces* bacterial (Thompson et al. 1987), has been widely used to generate transgenic varieties of crops including canola, cotton, maize etc., for resistance to glufosinate which interferes with the biosynthesis of the amino acid glutamine and ammonia detoxification and causes cessation of photosynthesis. Therefore, this approach has the great potential to identify desirable genes to improve the agronomic trait of crop varieties. Figure 6 shows the basic scheme of FOX hunting system: Full length cDNAs were cloned into the FOX hunting over-expression vectors. Those individual vectors were transformed into host plants by mixed *Agrobacterium*. The full length cDNA over-expression cassettes were randomly inserted into the host genome and the ectopic expression of exogenous full length cDNAs can induce abnormal phenotypes.

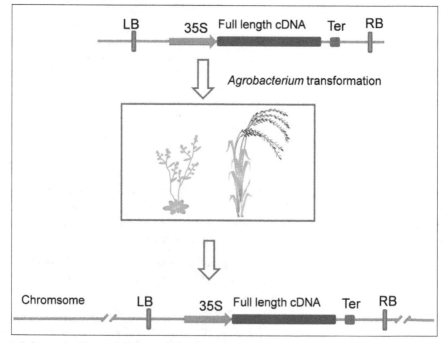

Fig. 6. Scheme for the generation of FOX hunting transgenic plants.

4. Summary of the transgenic resources

By reviewing the research progress in past few years, we are always impressed by that most of the basic scientific discoveries in plant field were using *Arabidopsis* or rice as the model plants. One of the significant reasons is that these two species are easiest to be transformed than other plants and thus the researchers can take the advantages of the public available transgenic resources to test their hypothesis in a relatively short period of time. In the following sections, we summarized the transgenic resources in Table 1 and described those representative ones for each of the methods. Some of the MGE insertion mutant lines are generated by endogenous transposon or retrotransposon rather than artificial transgene. Regarding those MGEs disrupt the gene functions in the similar way as T-DNA insertion and MGE mutant lines represent an important portion of genetics resources, we also introduce them briefly in this review for the interest of readers.

4.1 T-DNA insertion resources

SALK T-DNA insertion database is the most successful example in the history of transgenic resource so far. In 2003, over 225,000 *Arabidopsis* T-DNA insertion mutants have been generated by the SALK Institute, resulting in more than 88,000 insertions with precise locations determined in the whole genome of *Arabidopsis* (Alonso et al. 2003). Till recently, the numbers of T-DNA insert sites mapped to the genome of *Arabidopsis* by the Salk Institute have been over 150,000. Combined with the collections of other resources (SAIL, Wisc and GABI-KAT) around the world (Sessions et al. 2002; Rosso et al. 2003; Nishal et al. 2005; Li et al. 2007), totally 25,762 genes were identified with at least one T-DNA insertion mutation, which represent nearly 83% of the entire 31,128 protein-coding and non-coding RNA genes in the *Arabidopsis* genome (http://natural.salk.edu/geno/sum.txt). In addition, large-scale genotyping of those T-DNA insertion mutant lines has been ongoing for nearly 6 years to obtain as much as possible homozygous insertion mutants. As of today, 44,122 homozygote T-DNA lines, representing 24,476 individual genes have been sent to *Arabidopsis* Biological Resource Center (ABRC) for reproduction and distribution (http://signal.salk.edu/cgi-bin/homozygotes.cgi) (http://abrc.osu.edu/). The SALK Homozygote T-DNA Collection Project is nearly completed and this resource greatly facilitates the researchers to analyze the comprehensive phenotype and further understand the systematic gene functions at a genome wide level. Some other T-DNA insertion mutant resources are available for researchers to refer and utilize, which are listed in the Table 1.

4.2 Enhancer trap resources

The earlier Enhancer trap resource is reported in 1995 by Robert Martienssen and colleagues in the Cold Spring Harbor Laboratory (CSHL). They designed an ingenious strategy which combined the *Ac/Ds* MGE insertion method, Gene trap and Enhancer trap technology to generate a large DNA insertion population and 21,661 insertion events have been mapped to the genome of *Arabidopsis* (Sundaresan et al. 1995; Martienssen 1998). NASC established by Jim Haseloff and colleagues in the University of Cambridge contains 250 GAL4-GFP enhancer-trap lines which using GFP as report gene (Haseloff et al. 1997). Based on the similar strategy, the researchers in the Department of Biology, University of Pennsylvania established Enhancertraps database providing information of *Arabidopsis* lines transformed with Jim Haseloff's GAL4 enhancer trap vector. This database currently release 510 enhancer trap transgenic lines that show specific expression pattern in multiple kinds of organs, tissues or cell type. RMD (Rice Mutant Database) is a successful enhancer trap resource of rice generated by Qifa Zhang and colleagues as describe in section 2.2. Since first released in 2006, approximate 132,193 T-DNA insertion lines generated by this enhancer trap system have been characterized with the flanking sequence of insertion, report genes expression or phenotype of transgenic mutation. RMD currently releases the comprehensive information of the large scaled transgenic lines including the flanking sequences of T-DNA insertion sites, seed availability, reporter-gene expression patterns, as well as mutant phenotypes, etc (http://rmd.ncpgr.cn/).

4.3 MGE insertion resources

Multiple MGEs identified in maize and some of them have been exploited to generate insertion mutant resources for maize as well as other plants such as *Arabidopsis*. For example, Exon Trapping Insert Consortium (EXOTIC) utilized both *Ac/Ds* system and *Enhancer/Suppressor-mutator* (*En/Spm*) system to create large insertion mutant populations and 23,537 insertion sites have been mapped to the genome of *Arabidopsis* (Tissier et al. 1999). Also using the *Ds* element, RIKEN BioResource Center (RIKEN BRC) built a resource containing 18,566 *Arabidopsis* mutants with insertion positions mapped to the genome. As an alternative, *tos17* retrotransposon characterized in rice is employed in mutational analysis of rice genome by several institutes such as National Institute of Agrobiological Sciences (NIAS)(Miyao et al. 2003) and Centre de coopération internationale en recherche agronomique pour le développement (CIRAD)(Sallaud et al. 2004). Because other plant species are not easy to be transformed by direct trangene method, MGE insertion may become a more and more important strategy for the functional genomics studies of those plants other than *Arabidopsis* or rice.

4.4 Activation tagging resources

As a useful supplement of loss-of-function mutant resources mentioned above, activation tagging technique was developed and widely used to generate gain-of-function resources. Currently, there are 22,600 activation-tagging lines available in SALK and NASC (Weigel et al. 2000). RIKEN also established a large scale of activation-tagging lines with the number of 32,650 (Nakazawa et al. 2003). With the same method, 47,932 T-DNA tag lines in japonica rice were generated using activation-tagging vectors in POSTECH Rice T-DNA Insertion Sequence Database (RISD) (Jeong et al. 2006). Taiwan Rice Insertional Mutant (TRIM) created about 45,000 activation tagging lines using a T-DNA vector containing an enhancer octamer(Hsing et al. 2007). As a T-DNA insertion resource that using hosts plants other than *Arabidopsis* and rice, GmGenesDB in University of Missouri created 900 activation tagging lines of soybean.

4.5 Fox hunting resources

FOX hunting system was developed by a novel gain-of-function system that used normalized full-length cDNA introduced into host plants via large scaled transformation (Ichikawa et al. 2006). The first FOX hunting resource were established by transformation of

Arabidopsis with 10,000 *Arabidopsis* full length cDNA driven by 35S promoter, which contains more than 15,000 transgenic lines with 2.6 cDNA insertions on average in the genome. By a similar approach, they created 12,000 independent FOX hunting lines by transformation of rice with rice full length cDNA (Nakamura et al. 2007). The third FOX hunting system were created by transformation of *Arabidopsis* with the rice Full length cDNA for generating a heterologous gene resource which contain more than 23,000 independent *Arabidopsis* transgenic lines that expressed rice fl-cDNAs (Kondou et al. 2009).

5. Perspective

Take advantage of the dramatic improvements in transgenic efficiency, researchers are able to generate gain-of-function and loss-of-function transgenic resources with vast numbers of transformed *Arabidopsis* and rice plants. As a general procedure to utilize the loss-of-function resources, the individual researchers firstly order their interested insertion mutant lines by web based search of related database and then perform a PCR based genotyping to confirm the T-DNA insertion in their interested gene. Following a careful analysis of phenotype of the insertion homozygote lines, the researchers will confirm the abnormal phenotype of the mutants caused by T-DNA insertion via complementary experiment which is to transform the insertion mutant with a functional intact gene to rescue the phenotype. One of the common problems faced to the researches is that how should we do if the homozygote mutants show no obviously phenotype due to homolog gene redundancy or redundant genetic pathway to other gene with similar function. The first solution is to order all the redundant mutants and cross them to create double, triple or even quadruple mutant. If the redundant genes are correlated too close on the chromosome, it would be impossible to get the homozygote multiple mutants. Fortunately, other methods such as RNA interference, artificial micro RNA technology or TILLING (Targeting Induced Local Lesions in Genome) method are developed which can combine with the genetic cross method to create loss-of-function mutant of multiple gene loci (Schwab et al. 2006; Till et al. 2007; Warthmann et al. 2008). However, genetic pathway redundancies rather than homolog gene redundancies are not rare. In such case, the researchers will have neither homolog genes nor literature information to help them to figure out which gene loci should they use to create multiple mutant. New strategy may be developed to generate multiple site mutation resources for the researchers to screen obvious phenotype and further digest the complex genetic pathway.

To compensate the shortage of loss-of-function methods, gain-of-function transgenic can cause obvious phenotype even if the gene is redundant with other gene or the loss-of-function insertion is lethal. In addition, gain-of-function method has potential to discover useful genes which can be utilized to improve the agronomic traits of economic plants. For example, salt cress (*Thellungiella halophila*) is a very salt-tolerant species which are closely related to *Arabidopsis* (90-95% DNA sequence identity). In order to isolate salt stress tolerance genes, salt cress cDNAs under CaMV 35S promoter were transformed into *Arabidopsis* in a large scale and two genes ST6-66 and ST225 were discovered to improve the salt tolerance of *Arabidopsis* (Du et al. 2008). As another example described before, *Arabidopsis* FOX lines with ectopic over expression of rice specific genes in *Arabidopsis* also show abnormal phenotypes, further indicating that genes from various germplasms could introduce novo functions. Actually, genes from distant species have already been successfully used to improve the agronomic traits of crops such as *BAR* gene used for weed control, or *CRY* gene (encoding Bt toxin in *Bacillus thuringiensis*) used for pests control (Thompson et al. 1987; Bravo et al. 2007). We speculate that gain-of-function methods will be further engaged to identify more useful genes to improve crop traits as an important goal of plant science.

	method	Resource	Host plant	Web site for the resource	Reference
Lose of Function	T-DNA insertion	SALK Institute, ABRC	*Arabidopsis*	http://signal.salk.edu/tabout.html	(Alonso et al. 2003)
		Syngenta, SAIL	*Arabidopsis*	N/A	(Sessions et al. 2002)
		WiscDslox	*Arabidopsis*	http://www.hort.wisc.edu	(Nishal et al. 2005)
		GABI	*Arabidopsis*	http://www.gabi-kat.de	(Rosso et al. 2003; Li et al. 2007)
		INRA, FLAGdb	*Arabidopsis*	http://urgv.evry.inra.fr/FLAGdb	(Samson et al. 2002)
		POSTECH	Rice	http://www.postech.ac.kr/life/pfg/risd/index.htm	(Jeon et al. 2000)
		SHIP	Rice	http://ship.plantsignal.cn/index.do	(Fu et al. 2009)
		TRIM	Rice	http://trim.sinica.edu.tw/	(Hsing et al. 2007)
		ZJU	Rice	http://www.genomics.zju.edu.cn/ricetdna.html	(Chen et al. 2003)
	Enhancer trap	CSHL	*Arabidopsis*	http://genetrap.cshl.org/	(Sundaresan et al. 1995; Martienssen 1998)
		NASC	*Arabidopsis*	http://arabidopsis.info/CollectionInfo?id=24	n/a
		University of Penn	*Arabidopsis*	http://enhancertraps.bio.upenn.edu/default.html	n/a
		RDM	Rice	http://rmd.ncpgr.cn	(Zhang et al. 2006)
	MGE insertion	EXOTIC	*Arabidopsis*	http://www.jic.bbsrc.ac.uk/science/cdb/exotic/index.htm	(Tissier et al. 1999)
		RIKEN	*Arabidopsis*	http://rarge.gsc.riken.go.jp/dsmutant/index.pl	(Ito et al. 2002; Kuromori et al. 2004; Ito et al. 2005)
		NIAS	Rice	http://www.dna.affrc.go.jp/database	(Miyao et al. 2003)
		OTL, CIRAD	Rice	http://urgi.versailles.inra.fr/OryzaTagLine/	(Sallaud et al. 2004)
		UCD	Rice	http://www-plb.ucdavis.edu/labs/sundar/Rice_Genomics.htm	(Kolesnik et al. 2004)

		CSIRO	Rice	http://www.pi.csiro.au/fgrttpub/	(Eamens et al. 2004)
		GSNU	Rice	N/A	(Kim et al. 2004)
		EU-OSTID	Rice	http://orygenesdb.cirad.fr	(van Enckevort et al. 2005)
		Maize GDB	Maize	http://www.maizegdb.org/rescuemu-phenotype.php	(Fernandes et al. 2004)
Gain of function	Activation tagging	SALK, NASC	*Arabidopsis*	http://arabidopsis.info/CollectionInfo?id=59	(Weigel et al. 2000)
		RIKEN	*Arabidopsis*	http://activation.psc.database.riken.jp	(Nakazawa et al. 2003)
		Plant Research International	*Arabidopsis*	N/A	(Marsch-Martinez et al. 2002)
		TAMARA	*Arabidopsis*	http://arabidopsis.info/CollectionInfo?id=71	(Schneider et al. 2005)
		NI Vaviliv Institute of General Genetics RAS	*Arabidopsis*	N/A	(Pogorelko et al. 2008)
		JIC activate line	*Arabidopsis*	http://arabidopsis.info/CollectionInfo?id=29	n/a
		RISD	Rice	http://www.postech.ac.kr/life/pfg/risd	(Jeong et al. 2006)
		TRIM	Rice	http://trim.sinica.edu.tw	(Hsing et al. 2007)
		GmGenesDB	soybean	http://digbio.missouri.edu/gmgenedb/index.php	(Mathieu et al. 2009)
	FOX hunting	RIKEN	*Arabidopsis*	http:// nazunafox.psc.database.riken.jp	(Ichikawa et al. 2006)
		RIKEN, NIAS,RIBS Okayama	*Arabidopsis*	http:// ricefox.psc.riken.jp	(Kondou et al. 2009)
		NIAS	Rice	N/A	(Nakamura et al. 2007)

Table 1. T-DNA or MGE insertion mutant resources (N/A, not available).

6. Acknowledgment

Research in the authors' laboratories is supported by the National Natural Science Foundation of China (Grant No. 31171352) and National Transgenic Crop Initiative (Grant No. 2010ZX08010-002).

7. References

Alonso, J.M., Stepanova, A.N., Leisse, T.J., Kim, C.J., Chen, H., Shinn, P., Stevenson, D.K., Zimmerman, J., Barajas, P., Cheuk, R., Gadrinab, C., Heller, C., Jeske, A., Koesema, E., Meyers, C.C., Parker, H., Prednis, L., Ansari, Y., Choy, N., Deen, H., Geralt, M.,

Hazari, N., Hom, E., Karnes, M., Mulholland, C., Ndubaku, R., Schmidt, I., Guzman, P., Aguilar-Henonin, L., Schmid, M., Weigel, D., Carter, D.E., Marchand, T., Risseeuw, E., Brogden, D., Zeko, A., Crosby, W.L., Berry, C.C., and Ecker, J.R. 2003. Genome-wide insertional mutagenesis of Arabidopsis thaliana. *Science* 301(5633): 653-657.

An, G., Jeong, D.H., Jung, K.H., and Lee, S. 2005. Reverse genetic approaches for functional genomics of rice. *Plant Mol Biol* 59(1): 111-123.

Bechtold, N., Ellis, J. and Pelletier, G. 2003. In planta *Agrobacterium* mediated gene transfer by infiltration of adult Arabidopsis thanliana plants. *C R Acad Sci Paris, Life Sciences* 316: 1194-1199.

Bellen, H.J. 1999. Ten years of enhancer detection: lessons from the fly. *Plant Cell* 11(12): 2271-2281.

Bravo, A., Gill, S.S., and Soberon, M. 2007. Mode of action of Bacillus thuringiensis Cry and Cyt toxins and their potential for insect control. *Toxicon* 49(4): 423-435.

Campisi, L., Yang, Y., Yi, Y., Heilig, E., Herman, B., Cassista, A.J., Allen, D.W., Xiang, H., and Jack, T. 1999. Generation of enhancer trap lines in Arabidopsis and characterization of expression patterns in the inflorescence. *Plant J* 17(6): 699-707.

Casadaban, M.J. and Cohen, S.N. 1979. Lactose genes fused to exogenous promoters in one step using a Mu-lac bacteriophage: in vivo probe for transcriptional control sequences. *Proc Natl Acad Sci U S A* 76(9): 4530-4533.

Chen, S., Jin, W., Wang, M., Zhang, F., Zhou, J., Jia, Q., Wu, Y., Liu, F., and Wu, P. 2003. Distribution and characterization of over 1000 T-DNA tags in rice genome. *Plant J* 36(1): 105-113.

Clough, S.J. and Bent, A.F. 1998. Floral dip: a simplified method for *Agrobacterium*-mediated transformation of Arabidopsis thaliana. *Plant J* 16(6): 735-743.

Du, J., Huang, Y.P., Xi, J., Cao, M.J., Ni, W.S., Chen, X., Zhu, J.K., Oliver, D.J., and Xiang, C.B. 2008. Functional gene-mining for salt-tolerance genes with the power of Arabidopsis. *Plant J* 56(4): 653-664.

Eamens, A.L., Blanchard, C.L., Dennis, E.S., and Upadhyaya, N.M. 2004. A bidirectional gene trap construct suitable for T-DNA and Ds-mediated insertional mutagenesis in rice (Oryza sativa L.). *Plant Biotechnol J* 2(5): 367-380.

Fedoroff, N.V. and Smith, D.L. 1993. A versatile system for detecting transposition in Arabidopsis. *Plant J* 3(2): 273-289.

Fernandes, J., Dong, Q., Schneider, B., Morrow, D.J., Nan, G.L., Brendel, V., and Walbot, V. 2004. Genome-wide mutagenesis of Zea mays L. using RescueMu transposons. *Genome Biol* 5(10): R82.

Fu, F.F., Ye, R., Xu, S.P., and Xue, H.W. 2009. Studies on rice seed quality through analysis of a large-scale T-DNA insertion population. *Cell Res* 19(3): 380-391.

Giaever, G., Chu, A.M., Ni, L., Connelly, C., Riles, L., Veronneau, S., Dow, S., Lucau-Danila, A., Anderson, K., Andre, B., Arkin, A.P., Astromoff, A., El-Bakkoury, M., Bangham, R., Benito, R., Brachat, S., Campanaro, S., Curtiss, M., Davis, K., Deutschbauer, A., Entian, K.D., Flaherty, P., Foury, F., Garfinkel, D.J., Gerstein, M., Gotte, D., Guldener, U., Hegemann, J.H., Hempel, S., Herman, Z., Jaramillo, D.F., Kelly, D.E., Kelly, S.L., Kotter, P., LaBonte, D., Lamb, D.C., Lan, N., Liang, H., Liao, H., Liu, L., Luo, C., Lussier, M., Mao, R., Menard, P., Ooi, S.L., Revuelta, J.L., Roberts, C.J., Rose, M., Ross-Macdonald, P., Scherens, B., Schimmack, G., Shafer, B., Shoemaker, D.D., Sookhai-Mahadeo, S., Storms, R.K., Strathern, J.N., Valle, G., Voet, M.,

Volckaert, G., Wang, C.Y., Ward, T.R., Wilhelmy, J., Winzeler, E.A., Yang, Y., Yen, G., Youngman, E., Yu, K., Bussey, H., Boeke, J.D., Snyder, M., Philippsen, P., Davis, R.W., and Johnston, M. 2002. Functional profiling of the Saccharomyces cerevisiae genome. *Nature* 418(6896): 387-391.

Haseloff, J., Siemering, K.R., Prasher, D.C., and Hodge, S. 1997. Removal of a cryptic intron and subcellular localization of green fluorescent protein are required to mark transgenic Arabidopsis plants brightly. *Proc Natl Acad Sci U S A* 94(6): 2122-2127.

Hirochika, H. 2001. Contribution of the Tos17 retrotransposon to rice functional genomics. *Curr Opin Plant Biol* 4(2): 118-122.

Hirochika, H., Sugimoto, K., Otsuki, Y., Tsugawa, H., and Kanda, M. 1996. Retrotransposons of rice involved in mutations induced by tissue culture. *Proc Natl Acad Sci U S A* 93(15): 7783-7788.

Hiwatashi, Y., Nishiyama, T., Fujita, T., and Hasebe, M. 2001. Establishment of gene-trap and enhancer-trap systems in the moss Physcomitrella patens. *Plant J* 28(1): 105-116.

Hsing, Y.I., Chern, C.G., Fan, M.J., Lu, P.C., Chen, K.T., Lo, S.F., Sun, P.K., Ho, S.L., Lee, K.W., Wang, Y.C., Huang, W.L., Ko, S.S., Chen, S., Chen, J.L., Chung, C.I., Lin, Y.C., Hour, A.L., Wang, Y.W., Chang, Y.C., Tsai, M.W., Lin, Y.S., Chen, Y.C., Yen, H.M., Li, C.P., Wey, C.K., Tseng, C.S., Lai, M.H., Huang, S.C., Chen, L.J., and Yu, S.M. 2007. A rice gene activation/knockout mutant resource for high throughput functional genomics. *Plant Mol Biol* 63(3): 351-364.

Ichikawa, T., Nakazawa, M., Kawashima, M., Iizumi, H., Kuroda, H., Kondou, Y., Tsuhara, Y., Suzuki, K., Ishikawa, A., Seki, M., Fujita, M., Motohashi, R., Nagata, N., Takagi, T., Shinozaki, K., and Matsui, M. 2006. The FOX hunting system: an alternative gain-of-function gene hunting technique. *Plant J* 48(6): 974-985.

Ito, T., Motohashi, R., Kuromori, T., Mizukado, S., Sakurai, T., Kanahara, H., Seki, M., and Shinozaki, K. 2002. A new resource of locally transposed Dissociation elements for screening gene-knockout lines in silico on the Arabidopsis genome. *Plant Physiol* 129(4): 1695-1699.

Ito, T., Motohashi, R., Kuromori, T., Noutoshi, Y., Seki, M., Kamiya, A., Mizukado, S., Sakurai, T., and Shinozaki, K. 2005. A resource of 5,814 dissociation transposon-tagged and sequence-indexed lines of Arabidopsis transposed from start loci on chromosome 5. *Plant Cell Physiol* 46(7): 1149-1153.

Jeon, J.S., Lee, S., Jung, K.H., Jun, S.H., Jeong, D.H., Lee, J., Kim, C., Jang, S., Yang, K., Nam, J., An, K., Han, M.J., Sung, R.J., Choi, H.S., Yu, J.H., Choi, J.H., Cho, S.Y., Cha, S.S., Kim, S.I., and An, G. 2000. T-DNA insertional mutagenesis for functional genomics in rice. *Plant J* 22(6): 561-570.

Jeong, D.H., An, S., Kang, H.G., Moon, S., Han, J.J., Park, S., Lee, H.S., An, K., and An, G. 2002. T-DNA insertional mutagenesis for activation tagging in rice. *Plant Physiol* 130(4): 1636-1644.

Jeong, D.H., An, S., Park, S., Kang, H.G., Park, G.G., Kim, S.R., Sim, J., Kim, Y.O., Kim, M.K., Kim, J., Shin, M., Jung, M., and An, G. 2006. Generation of a flanking sequence-tag database for activation-tagging lines in japonica rice. *Plant J* 45(1): 123-132.

Kim, C.M., Piao, H.L., Park, S.J., Chon, N.S., Je, B.I., Sun, B., Park, S.H., Park, J.Y., Lee, E.J., Kim, M.J., Chung, W.S., Lee, K.H., Lee, Y.S., Lee, J.J., Won, Y.J., Yi, G., Nam, M.H., Cha, Y.S., Yun, D.W., Eun, M.Y., and Han, C.D. 2004. Rapid, large-scale generation of Ds transposant lines and analysis of the Ds insertion sites in rice. *Plant J* 39(2): 252-263.

Kolesnik, T., Szeverenyi, I., Bachmann, D., Kumar, C.S., Jiang, S., Ramamoorthy, R., Cai, M., Ma, Z.G., Sundaresan, V., and Ramachandran, S. 2004. Establishing an efficient Ac/Ds tagging system in rice: large-scale analysis of Ds flanking sequences. *Plant J* 37(2): 301-314.

Kondou, Y., Higuchi, M., Takahashi, S., Sakurai, T., Ichikawa, T., Kuroda, H., Yoshizumi, T., Tsumoto, Y., Horii, Y., Kawashima, M., Hasegawa, Y., Kuriyama, T., Matsui, K., Kusano, M., Albinsky, D., Takahashi, H., Nakamura, Y., Suzuki, M., Sakakibara, H., Kojima, M., Akiyama, K., Kurotani, A., Seki, M., Fujita, M., Enju, A., Yokotani, N., Saitou, T., Ashidate, K., Fujimoto, N., Ishikawa, Y., Mori, Y., Nanba, R., Takata, K., Uno, K., Sugano, S., Natsuki, J., Dubouzet, J.G., Maeda, S., Ohtake, M., Mori, M., Oda, K., Takatsuji, H., Hirochika, H., and Matsui, M. 2009. Systematic approaches to using the FOX hunting system to identify useful rice genes. *Plant J* 57(5): 883-894.

Kuromori, T., Hirayama, T., Kiyosue, Y., Takabe, H., Mizukado, S., Sakurai, T., Akiyama, K., Kamiya, A., Ito, T., and Shinozaki, K. 2004. A collection of 11 800 single-copy Ds transposon insertion lines in Arabidopsis. *Plant J* 37(6): 897-905.

Li, Y., Rosso, M.G., Viehoever, P., and Weisshaar, B. 2007. GABI-Kat SimpleSearch: an Arabidopsis thaliana T-DNA mutant database with detailed information for confirmed insertions. *Nucleic Acids Res* 35(Database issue): D874-878.

Liu, Y.G., Mitsukawa, N., Oosumi, T., and Whittier, R.F. 1995. Efficient isolation and mapping of Arabidopsis thaliana T-DNA insert junctions by thermal asymmetric interlaced PCR. *Plant J* 8(3): 457-463.

Liu, Y.G. and Whittier, R.F. 1995. Thermal asymmetric interlaced PCR: automatable amplification and sequencing of insert end fragments from P1 and YAC clones for chromosome walking. *Genomics* 25(3): 674-681.

Marsch-Martinez, N., Greco, R., Van Arkel, G., Herrera-Estrella, L., and Pereira, A. 2002. Activation tagging using the En-I maize transposon system in Arabidopsis. *Plant Physiol* 129(4): 1544-1556.

Martienssen, R.A. 1998. Functional genomics: probing plant gene function and expression with transposons. *Proc Natl Acad Sci U S A* 95(5): 2021-2026.

Mathieu, M., Winters, E.K., Kong, F., Wan, J., Wang, S., Eckert, H., Luth, D., Paz, M., Donovan, C., Zhang, Z., Somers, D., Wang, K., Nguyen, H., Shoemaker, R.C., Stacey, G., and Clemente, T. 2009. Establishment of a soybean (Glycine max Merr. L) transposon-based mutagenesis repository. *Planta* 229(2): 279-289.

Mc, C.B. 1950. The origin and behavior of mutable loci in maize. *Proc Natl Acad Sci U S A* 36(6): 344-355.

Miyao, A., Iwasaki, Y., Kitano, H., Itoh, J., Maekawa, M., Murata, K., Yatou, O., Nagato, Y., and Hirochika, H. 2007. A large-scale collection of phenotypic data describing an insertional mutant population to facilitate functional analysis of rice genes. *Plant Mol Biol* 63(5): 625-635.

Miyao, A., Tanaka, K., Murata, K., Sawaki, H., Takeda, S., Abe, K., Shinozuka, Y., Onosato, K., and Hirochika, H. 2003. Target site specificity of the Tos17 retrotransposon shows a preference for insertion within genes and against insertion in retrotransposon-rich regions of the genome. *Plant Cell* 15(8): 1771-1780.

Nakamura, H., Hakata, M., Amano, K., Miyao, A., Toki, N., Kajikawa, M., Pang, J., Higashi, N., Ando, S., Toki, S., Fujita, M., Enju, A., Seki, M., Nakazawa, M., Ichikawa, T., Shinozaki, K., Matsui, M., Nagamura, Y., Hirochika, H., and Ichikawa, H. 2007. A genome-wide gain-of function analysis of rice genes using the FOX-hunting system. *Plant Mol Biol* 65(4): 357-371.

Nakazawa, M., Ichikawa, T., Ishikawa, A., Kobayashi, H., Tsuhara, Y., Kawashima, M., Suzuki, K., Muto, S., and Matsui, M. 2003. Activation tagging, a novel tool to dissect the functions of a gene family. *Plant J* 34(5): 741-750.

Nishal, B., Tantikanjana, T., and Sundaresan, V. 2005. An inducible targeted tagging system for localized saturation mutagenesis in Arabidopsis. *Plant Physiol* 137(1): 3-12.

Ostergaard, L. and Yanofsky, M.F. 2004. Establishing gene function by mutagenesis in Arabidopsis thaliana. *Plant J* 39(5): 682-696.

Parinov, S. and Sundaresan, V. 2000. Functional genomics in Arabidopsis: large-scale insertional mutagenesis complements the genome sequencing project. *Curr Opin Biotechnol* 11(2): 157-161.

Piffanelli, P., Droc, G., Mieulet, D., Lanau, N., Bes, M., Bourgeois, E., Rouviere, C., Gavory, F., Cruaud, C., Ghesquiere, A., and Guiderdoni, E. 2007. Large-scale characterization of Tos17 insertion sites in a rice T-DNA mutant library. *Plant Mol Biol* 65(5): 587-601.

Pogorelko, G.V., Fursova, O.V., Ogarkova, O.A., and Tarasov, V.A. 2008. A new technique for activation tagging in Arabidopsis. *Gene* 414(1-2): 67-75.

Ross-Macdonald, P., Coelho, P.S., Roemer, T., Agarwal, S., Kumar, A., Jansen, R., Cheung, K.H., Sheehan, A., Symoniatis, D., Umansky, L., Heidtman, M., Nelson, F.K., Iwasaki, H., Hager, K., Gerstein, M., Miller, P., Roeder, G.S., and Snyder, M. 1999. Large-scale analysis of the yeast genome by transposon tagging and gene disruption. *Nature* 402(6760): 413-418.

Rosso, M.G., Li, Y., Strizhov, N., Reiss, B., Dekker, K., and Weisshaar, B. 2003. An Arabidopsis thaliana T-DNA mutagenized population (GABI-Kat) for flanking sequence tag-based reverse genetics. *Plant Mol Biol* 53(1-2): 247-259.

Sallaud, C., Gay, C., Larmande, P., Bes, M., Piffanelli, P., Piegu, B., Droc, G., Regad, F., Bourgeois, E., Meynard, D., Perin, C., Sabau, X., Ghesquiere, A., Glaszmann, J.C., Delseny, M., and Guiderdoni, E. 2004. High throughput T-DNA insertion mutagenesis in rice: a first step towards in silico reverse genetics. *Plant J* 39(3): 450-464.

Samson, F., Brunaud, V., Balzergue, S., Dubreucq, B., Lepiniec, L., Pelletier, G., Caboche, M., and Lecharny, A. 2002. FLAGdb/FST: a database of mapped flanking insertion sites (FSTs) of Arabidopsis thaliana T-DNA transformants. *Nucleic Acids Res* 30(1): 94-97.

Sasaki, T. and Burr, B. 2000. International Rice Genome Sequencing Project: the effort to completely sequence the rice genome. *Curr Opin Plant Biol* 3(2): 138-141.

Schneider, A., Kirch, T., Gigolashvili, T., Mock, H.P., Sonnewald, U., Simon, R., Flugge, U.I., and Werr, W. 2005. A transposon-based activation-tagging population in Arabidopsis thaliana (TAMARA) and its application in the identification of dominant developmental and metabolic mutations. *FEBS Lett* 579(21): 4622-4628.

Schwab, R., Ossowski, S., Riester, M., Warthmann, N., and Weigel, D. 2006. Highly specific gene silencing by artificial microRNAs in Arabidopsis. *Plant Cell* 18(5): 1121-1133.

Sessions, A., Burke, E., Presting, G., Aux, G., McElver, J., Patton, D., Dietrich, B., Ho, P., Bacwaden, J., Ko, C., Clarke, J.D., Cotton, D., Bullis, D., Snell, J., Miguel, T., Hutchison, D., Kimmerly, B., Mitzel, T., Katagiri, F., Glazebrook, J., Law, M., and Goff, S.A. 2002. A high-throughput Arabidopsis reverse genetics system. *Plant Cell* 14(12): 2985-2994.

Spertini, D., Beliveau, C., and Bellemare, G. 1999. Screening of transgenic plants by amplification of unknown genomic DNA flanking T-DNA. *Biotechniques* 27(2): 308-314.

Sundaresan, V., Springer, P., Volpe, T., Haward, S., Jones, J.D., Dean, C., Ma, H., and Martienssen, R. 1995. Patterns of gene action in plant development revealed by enhancer trap and gene trap transposable elements. *Genes Dev* 9(14): 1797-1810.

Thompson, C.J., Movva, N.R., Tizard, R., Crameri, R., Davies, J.E., Lauwereys, M., and Botterman, J. 1987. Characterization of the herbicide-resistance gene bar from Streptomyces hygroscopicus. *EMBO J* 6(9): 2519-2523.

Till, B.J., Cooper, J., Tai, T.H., Colowit, P., Greene, E.A., Henikoff, S., and Comai, L. 2007. Discovery of chemically induced mutations in rice by TILLING. *BMC Plant Biol* 7: 19.

Tissier, A.F., Marillonnet, S., Klimyuk, V., Patel, K., Torres, M.A., Murphy, G., and Jones, J.D. 1999. Multiple independent defective suppressor-mutator transposon insertions in Arabidopsis: a tool for functional genomics. *Plant Cell* 11(10): 1841-1852.

Utley, R.T., Ikeda, K., Grant, P.A., Cote, J., Steger, D.J., Eberharter, A., John, S., and Workman, J.L. 1998. Transcriptional activators direct histone acetyltransferase complexes to nucleosomes. *Nature* 394(6692): 498-502.

van Enckevort, L.J., Droc, G., Piffanelli, P., Greco, R., Gagneur, C., Weber, C., Gonzalez, V.M., Cabot, P., Fornara, F., Berri, S., Miro, B., Lan, P., Rafel, M., Capell, T., Puigdomenech, P., Ouwerkerk, P.B., Meijer, A.H., Pe, E., Colombo, L., Christou, P., Guiderdoni, E., and Pereira, A. 2005. EU-OSTID: a collection of transposon insertional mutants for functional genomics in rice. *Plant Mol Biol* 59(1): 99-110.

Wan, S., Wu, J., Zhang, Z., Sun, X., Lv, Y., Gao, C., Ning, Y., Ma, J., Guo, Y., Zhang, Q., Zheng, X., Zhang, C., Ma, Z., and Lu, T. 2009. Activation tagging, an efficient tool for functional analysis of the rice genome. *Plant Mol Biol* 69(1-2): 69-80.

Warthmann, N., Chen, H., Ossowski, S., Weigel, D., and Herve, P. 2008. Highly specific gene silencing by artificial miRNAs in rice. *PLoS One* 3(3): e1829.

Weigel, D., Ahn, J.H., Blazquez, M.A., Borevitz, J.O., Christensen, S.K., Fankhauser, C., Ferrandiz, C., Kardailsky, I., Malancharuvil, E.J., Neff, M.M., Nguyen, J.T., Sato, S., Wang, Z.Y., Xia, Y., Dixon, R.A., Harrison, M.J., Lamb, C.J., Yanofsky, M.F., and Chory, J. 2000. Activation tagging in Arabidopsis. *Plant Physiol* 122(4): 1003-1013.

Winzeler, E.A., Shoemaker, D.D., Astromoff, A., Liang, H., Anderson, K., Andre, B., Bangham, R., Benito, R., Boeke, J.D., Bussey, H., Chu, A.M., Connelly, C., Davis, K., Dietrich, F., Dow, S.W., El Bakkoury, M., Foury, F., Friend, S.H., Gentalen, E., Giaever, G., Hegemann, J.H., Jones, T., Laub, M., Liao, H., Liebundguth, N., Lockhart, D.J., Lucau-Danila, A., Lussier, M., M'Rabet, N., Menard, P., Mittmann, M., Pai, C., Rebischung, C., Revuelta, J.L., Riles, L., Roberts, C.J., Ross-MacDonald, P., Scherens, B., Snyder, M., Sookhai-Mahadeo, S., Storms, R.K., Veronneau, S., Voet, M., Volckaert, G., Ward, T.R., Wysocki, R., Yen, G.S., Yu, K., Zimmermann, K., Philippsen, P., Johnston, M., and Davis, R.W. 1999. Functional characterization of the S. cerevisiae genome by gene deletion and parallel analysis. *Science* 285(5429): 901-906.

Yamamoto, Y.Y., Tsuhara, Y., Gohda, K., Suzuki, K., and Matsui, M. 2003. Gene trapping of the Arabidopsis genome with a firefly luciferase reporter. *Plant J* 35(2): 273-283.

Zhang, J., Li, C., Wu, C., Xiong, L., Chen, G., Zhang, Q., and Wang, S. 2006. RMD: a rice mutant database for functional analysis of the rice genome. *Nucleic Acids Res* 34(Database issue): D745-748.

Transgenic Plants as a Tool
for Plant Functional Genomics

Inna Abdeeva[1], Rustam Abdeev[2],
Sergey Bruskin[1] and Eleonora Piruzian[1]
[1]NI Vavilov Institute of General Genetics RAS, Moscow
[2]Center for Theoretical Problems of Physico-Chemical Pharmacology RAS, Moscow
Russsia

1. Introduction

Functional genomics aim to discover the biological function of particular genes and to uncover how sets of genes and their products work together. Transgenic plants are proving to be powerful tools to study various aspects of plant sciences. The emerging scientific revolution sparked by genomics based technologies is producing enormous amounts of DNA sequence information that, together with plant transformation methodology, is opening up new experimental opportunities for functional genomics analysis.

2. Plant functional genomics methods

The main methods of Plant Functional Genomics are as follows.

2.1 Functional annotations for genes

Gene function prediction is based on comparison of genomes and proteomes with searching homologies within different species to gene of interest with known functions from nucleotide and amino acid databases . Putative genes can be identified by scanning a genome for regions likely to encode proteins, based on characteristics such as long open reading frames, transcriptional initiation sequences, and polyadenylation sites. A sequence identified as a putative gene must be confirmed by further evidence, such as similarity to cDNA or EST sequences from the same organism, similarity of the predicted protein sequence to known proteins, association with promoter sequences, or evidence that mutating the sequence produces an observable phenotype.

2.2 Gene-targeted and site-directed mutagenesis. Reverse genetics methods (loss of function)

Using transgenic plant with insertion/deletion or site-specific mutations. Host gene is replaced with mutant allele. The most conventional approach to the analysis of gene function is loss-of-function mutagenesis by chemicals or fast neutrons that introduce random mutations or deletions in the genome (Ostergaard and Yanofsky 2004).

Transferred DNA (T-DNA) tagging or transposon tagging methods were developed to generate loss-of-function mutations because these tag sequences can be used to identify the genes disrupted by these elements (Sundaresan and Ramachandran 2001; Sussman et al. 1999). However, because many plant genes in *Arabidopsis*, rice, and other plants belong to gene families (Goff et al. 2002; Kaul et al. 2000), the characterization of gene functions by single-gene mutagenesis is not always possible. Many mutants generated by single-gene disruption do not show clear phenotypes because of genetic redundancy.

2.3 Overexpression of normal gene in transgenic plants (gain of function)

Gain-of-function approaches have been used as an alternative or complementary method to loss-of-function approaches as well as to confer new functions to plants. Gain-of-function is achieved by increasing gene expression levels through the random activation of endogenous genes by transcriptional enhancers or the expression of individual transgenes by transformation. Gain-of-function mutagenesis is based on the random insertion of transcriptional enhancers into the genome or the expression of transgenes under the control of a strong promoter (Matsui et al. 2006; Nakazawa et al. 2003; Weigel et al. 2000) . In this approach, phenotypes of gain-of-function mutants that overexpress a member of a gene family can be observed without interference from other family members, which allows the characterization of functionally redundant genes(T. Ito and Meyerowitz 2000; Nakazawa et al. 2001).

Alternatively it is possible to overexpress mutant forms of a gene that interfere with the (wildtype) genes function. Over expression of a mutant gene may result in high levels of a non-functional protein resulting in a dominant negative interaction with the wild type protein. In this case the mutant version will out compete for the wild type proteins partners resulting in a mutant phenotype.

The advantages of gain-of function approaches in comparison to loss-of-function for the characterization of gene functions include the abilities to (*a*) analyze individual gene family members, (*b*) characterize the function of genes from nonmodel plants using a heterologous expression system, and (*c*) identify genes that confer stress tolerance to plants that result from the introduction of transgenes.

The first gain-of-function approach in plants was the activation-tagging system (Kakimoto 1996). In this system, T-DNA that harbors strong promoter or enhancer elements is randomly integrated into the plant genome. The introduced promoter or enhancer elements activate genes near the site of insertion.

Other recently developed gain-of-function approaches include cDNA overexpression and open reading frame (ORF) overexpression systems. In these approaches, cDNAs from Cdna libraries, representative full-length cDNAs (fl-cDNAs), or ORFs are strongly expressed when they were cloned downstream of a strong promoter.

The production of a large population of gain-of-function mutants can accelerate the high-throughput screening of desired mutants and the characterization of gene functions.

In the activation-tagging method, plant genes are randomly activated to produce gain-of function mutants. In this strategy, the promoter or enhancer elements from the *cauliflower mosaic virus* (*CaMV*) *35S* gene have been exploited (Odell et al. 1985). Genes near the insertion site are activated under the control of enhancer elements.

After the selection of mutants from the population of transformants, T-DNA insertion sites are determined to identify candidate genes.

Plasmid rescue, inverse PCR, or adapter PCR methods can be used to recover the genomic fragments near the T-DNA right and left border sequences (Spertini et al. 1999; Yamamoto et al. 2003). TAIL-PCR is also an efficient method to determine T-DNA insertion sites (Singer and Burke 2003)

Although many algorithms have been developed to predict the presence of transcriptional units within the genome, the accuracy of such predictions is still limited. Empirical information is required to correct annotation, and the main source of experimental information to achieve this is derived from RNA transcript analysis. Notable progress in *Arabidopsis* genome annotation has been made by the generation of expressed sequence tags, fl-cDNAs (Seki et al. 2002), and whole genome tiling array studies (Toyoda and Shinozaki 2005). The fl-cDNAs are also important as a resource for functional genomics, i.e., in the identification of gene function, because they contain all the information needed for the production of functional RNAs and proteins.

Approximately 240,000 *Arabidopsis* fl-cDNA clones have been generated (Sakurai et al. 2005; Seki et al. 2002) using the biotinylated CAP trapper method together with trehalose-thermoactivated reverse transcriptase (Carninci et al. 1996; Carninci et al. 1997; Carninci et al. 1998) . Large sets of fl-cDNA clones have also been produced from several plants such as rice (Kikuchi et al. 2003), wheat (Ogihara et al. 2004), poplar (Nanjo et al. 2007), soybean (Umezawa et al. 2008), barley (K. Sato et al. 2009), cassava (Sakurai et al. 2007), sitka spruce (Ralph et al. 2008), *Physcomitrella patens* (Nishiyama et al. 2003), and *Thellungiella halophila* (Taji et al. 2008).Well-characterized collections of cDNAs play an essential role in defining the function of genes and proteins in plants. The cDNA overexpression system is one of the approaches that use cDNA resources.

Progress in sequencing technology has revealed the genome sequences of many plant species that include *Arabidopsis*, rice, poplar, grape, papaya, and sorghum (2000; Jaillon et al. 2007; Ming et al. 2008; Paterson et al. 2009; Sasaki et al. 2002; Tuskan et al. 2006). A functional genomics approach is now required to clarify the function of genes in these plant species.

However, transgenic approaches for both forward and reverse genetic studies are not yet practical in many plants in which transformation methodology is inefficient or not available. A heterologous expression approach provides a solution for the high-throughput characterization of gene functions in these plant species.

One study used approximately 10,000 nonredundant fl-cDNA clones from the RIKEN *Arabidopsis* fl-cDNA collection (Seki et al. 2002). A representative of each flcDNA was mixed at approximately the same molar ratio to generate a cDNA mixture and then cloned into an expression vector under the control of the *CaMV 35S* promoter. This flcDNA expression library was used to transform *Arabidopsis* plants by *in planta* transformation. In these transgenic plants, fl-cDNAs are randomly expressed in individual *Arabidopsis* plants so that each plant carries one (or more) fl-cDNA(s).

The introduced fl-cDNAs can be cloned easily using vector-specific primers after the isolation of mutants. Thus, the cDNA that caused the mutant phenotype can be directly linked to a function.

The full-length cDNA over-expressing gene (FOX) hunting system is an alternative gain of-function approach that uses fl-cDNAs. The FOX hunting system was applied for the high-throughput analysis of rice genes by heterologous expression in *Arabidopsis* (Matsui et al. 2009). The efficient, rapid, and high-throughput transformation system developed in *Arabidopsis*, together with the short generation time and compact size of this plant, makes *Arabidopsis* an ideal host plant.

These advantages have enabled researchers to express heterologous genes in *Arabidopsis* to analyze their functions.

Whole genome sequencing makes it possible to predict the presence of genes in the genome. Of particular interest are the gene elements that encode proteins, called ORFs. Because ORFs can be distinguished from fl-cDNAs by their lack of 5_ and 3_ untranslated region (UTR) sequences, they can be considered a minimal unit of the gene that encodes information on the functional protein. The *Saccharomyces cerevisiae* ORFeome project was the first attempt to verify the genome annotation at the genomic scale and to clone all its predicted ORFs (Heyman et al. 1999). The ORF collection has been created for functional analysis in various organisms, e.g., RNAi approaches in *Caenorhabditis elegans* (Piano et al. 2005), cellular localization studies of YFP/GFP fusion proteins in *Schizosaccharomyces pombe* (Matsuyama et al. 2006), GFP fusion proteins in *Escherichia coli* (Kitagawa et al. 2005), and proteomics in human (Collins et al. 2004; Rual et al. 2004). These ORF clone collections can facilitate the large-scale analysis of individual genes.

2.4 Studying gene expression using DNA-RNA hybridization, gene silencing

Transgene expression in pair with reporter gene under control of inducible promoter allows to reveal temporal functional effects of gene expression and the compartmentalization of transgene products. The gene silencing techniques (also known as RNA-interference) allow to achieve temporary disrupting effects of gene expression (gene knockdown) . This procedure offer the possibility to explore gene expression more precise.

Because transgene-induced RNAi has been effective at silencing one or more genes in a wide range of plants, this technology also bears potential as a powerful functional genomics tool across the plant kingdom.

RNA-induced gene silencing (RNAi), was originally observed as unusual expression patterns of a transgene designed to induce overexpression of chalcone synthase in petunia plants (Napoli et al. 1990).

In the years following this observation, experiments in many model systems contributed to rapid advancements in understanding the underlying mechanisms, and RNA-mediated gene silencing processes came to be collectively known as RNA interference (RNAi). It is known that the 'triggers' for RNAi were small RNAs, 21–25 nts in length, that were processed from longer, double-stranded (ds) RNAs by endonuclease proteins referred to as dicers (Fire et al. 1998; Hamilton and Baulcombe 1999; Mello et al. 2001; Zamore et al. 2000).

These siRNAs cause direct degradation of mRNAs in a homology dependent manner and lead to post-transcriptional silencing of the silencing target. Other gene silencing methods are direct heterochromatin formation and DNA methylation at regulatory sequences for the target to be silenced, which alsoinduce transcriptional silencing of target loci in a homology dependent fashion (reviewed by ref. (Eamens et al. 2008). Now, it is understood that RNAi is an evolutionarily conserved mechanism for gene regulation that is critical for many examples of growth and development.

There are multiple pathways by which small RNA molecules can influence gene expression in plants, at both the transcriptional and post-transcriptional levels. These pathways vary in their sources of small RNAs and specific mechanisms of silencing (S. W. L. Chan 2008; Eamens et al. 2008; Verdel et al. 2009).

Because transgene-induced RNAi has been effective at silencing one or more genes in a wide range of plants, this technology also bears potential as a powerful functional genomics tool across the plant kingdom. A common strategy for functional genomics projects is to generate lines that are deficient for the activity of a subset of genes, and test the knock down lines for phenotypes to characterize the function of the knocked down gene.

In many cases, a single inverted repeat transgene can be designed to silence multiple, closely related genes (Springer et al. 2007).

To induce transcriptional silencing with a transgene, a typical strategy involves designing a construct such that a dsRNA is generated which bears homology to the promoter region of the intended silencing target (Mette et al. 2000). Herein, this method of silencing will be referred to as promoter directed RNA silencing.

To induce post-transcriptional silencing with a transgene, a portion of the coding region of the gene is typically introduced into an inverted repeat (IR) construct, and expression of that transgene will result in a dsRNA with homology to the coding region of the intended silencing target (McGinnis et al. 2005). This type of silencing is likely mediated by components of the trans-acting siRNA pathway in plants (reviewed by ref. Verdel et al. 2009). Herein, this method of silencing will be referred to as coding region directed RNA silencing.

2.5 Analysis of spatial and temporal expression of studied gene

Genomic studies tend to be done at the whole tissue/organ level due to the ease of collecting samples and/or the lack of tools necessary to isolate sufficient quantities of specific cell or tissue types. Recent studies, however, have shown that most transcriptional responses to environmental stimuli are cell-type specific (Dinneny 2008; Gifford et al. 2008). In addition, the many examples of ion-channels, hormone biosynthetic enzymes and signaling components with spatially complex expression patterns clearly illustrate the need to study all aspects of plant biology at high-spatial and temporal resolution to fully understand the plant–environment interaction.

The root of *Arabidopsis* provides an excellent system for generating and utilizing such tools due to the simple and stereotypical organization of tissues and cell types. Specific cell layers in the *Arabidopsis* root have been engineered to express green fluorescent protein (GFP). Fluorescence- activated cell sorting (FACS) can then be used to enrich for GFP-positive cells

(Birnbaum et al. 2003; Birnbaum et al. 2005). This method has been used to characterize the global transcriptional profiles of nearly all cell types in roots grown under standard conditions (Birnbaum et al. 2003; Brady et al. 2007) and to characterize transcriptional changes that occur in these cell types in response to salt stress, iron deprivation and nitrogen treatment (Dinneny et al. 2008; Gifford et al. 2008). A detailed description of these studies can be found in the following reviews of Dinneny (2010) and Iyer-Pascuzzi and co-workers (2009).

Genetically-encoded fluorophores offer a vast tool kit to study *in vivo* molecular events such as protein localization and gene expression. Fluorescent proteins have also been engineered to act as biosensors, which either emit fluorescence in response to a specific biological stimulus or undergo a change in intrinsic fluorescence intensity (Frommer et al. 2009).

Transgene expression is usually driven by a constitutive promoter. Thus, high expression levels in inappropriate tissue or developmental contexts might occur. This misexpression can cause the ectopic expression of endogenous genes and might result in a phenotype that is not related to the authentic functions of the transgene. In some cases, this misexpression can lead to the incorrect functional annotation of genes. Tissue-specific expression can provide information on intracellular events in each tissue. Replacing the *CaMV 35S* promoter with tissue-specific promoters is another way to analyze gene function in certain tissues.

Two component systems have been developed for conditional gene activation or silencing (Brand et al. 2006). They combine an activator locus that codes for an artificial transcription factor expressed in restricted tissues at precise developmental times.

The activation or the ectopic expression of developmentally controlled transcription factors sometimes causes an embryonic or seedling lethal phenotype, making it difficult to analyze the function of the gene . Thus, controlled gene expression by an inducible system might be an efficient approach to identify these genes (Zuo et al. 2002).

2.6 Microarrays

Microarrays allow the identification of candidate genes involved in a given process based on variation between transcript levels for different conditions and shared expression patterns with genes of known function. With appropriate controls and repeated experiments, significant data are obtained on gene expression profiles under various conditions (including stresses) or in various organs. Because of the large quantity of data produced by these techniques and the desire to find biologically meaningful patterns, bioinformatics is crucial to analyze functional genomics data. However, the DNA microarray and bioinformatics data are not sufficient for determining correct expression profiles due to limited accuracy of the obtained data. Next stage of investigations explores the properties and functions of selected genes. In this case, a transgenic plant construction is one of the most informative techniques.

2.7 Next generation sequencing

Previously, DNA sequencing was performed almost exclusively by the Sanger method, which has excellent accuracy and reasonable read length but very low throughput. Sanger sequencing was used to obtain the first sequence of the human genome in 2001 (Lander et al. 2001; Venter et al. 2001). Shortly thereafter, the second complete individual genome (James

D. Watson) was sequenced using next-generation technology, which marked the first human genome sequenced with new Next Generation Sequencing (NGS) technology (Wheeler et al. 2008). A common strategy for NGS is to use DNA synthesis or ligation process to read through many different DNA templates in parallel (Fuller et al. 2009). Therefore, NGS reads DNA templates in a highly parallel manner to generate massive amounts of sequencing data but, as mentioned above, the read length for each DNA template is relatively short (35–500 bp) compared to traditional Sanger sequencing (1000–1200 bp). NGS technologies have increased the speed and throughput capacities of DNA sequencing and, as a result, dramatically reduced overall sequencing costs (Metzker 2010).

Current NGS approaches can be classified into three major categories:

1. DNA-Seq. Genome-based sequencing yielding genomic deletions and rearrangements, copy-number variations (CNV) of smaller regions or elements, and single-nucleotide polymorphisms (SNPs).
2. RNA-Seq. RNA-Sequencing, yielding genome-wide and quantitative information about transcribed regions (exons, and subsequently transcripts).
3. Chromatin-immunoprecipitation (ChIP)-Seq. a) transcription factor (TF)-based ChIP, yielding genome-wide information about the physical binding sites of individual TFs to within a few hundred base pairs. b) Epigenetic ChIP (DNA methylation and/or histone modifications), yielding information about modifications and the accessibility of genomic regions to TFs and other factors.

The inclusion of NGS-based transcriptome sequencing for ChIP of transcription factor binding and epigenetic analyses (usually based on DNA methylation or histone modification ChIP) completes the picture with unprecedented resolution enabling the detection of even subtle differences such as alternative splicing of individual exons.

Next-generation sequencing technologies have found broad applicability in functional genomics research. Their applications in the field have included gene expression profiling, genome annotation, small non-coding RNA (ncRNA) discovery and profiling, and detection of aberrant transcription, which are areas that have been previously dominated by microarrays. Thus, functional genomics and systems biology approaches will benefit from the enormous data density intrinsic to NGS applications, which will beyond doubt play an important role both in definition as well as verification of mathematical models of biological systems such as a cell or a tissue.

As mentioned above the inventory of methods used to study gene product functions *in vivo* (i.e. in a living organism) includes gene silencing, induced mutagenesis, reporter gene strategy, microarrays, and some others. However, there are some limitations inherent to this type of approach. First of all, physiologically essential genes cannot be switched off, and the induced mutagenesis can lead to concomitant mutations. The use of microarrays can lead to misinterpretation of the results since changes in transcription are not always accompanied by changes in protein level (Mittler et al. 1998). Moreover, the transcription level fails to reflect post-translation modifications of protein products which often occur *in vivo*. It is also worth to mention that when an enzyme possesses many isoforms, it is difficult to measure the activity of each of them *in vivo* (Slakeski et al. 1990). In view of the above-mentioned limitations, development of novel models for functional genetics which will aid to overcome these difficulties is deemed very much desirable.

One of such models may be the approach that is developed in our laboratories that employ transgenic plants that constitutively express bacterial genes, which code enzymes that are functionally homologous to plant enzymes. Such an approach was proposed and used in our laboratory since mid-1980s (Piruzian et al. 1983; Piruzian and Andrianov 1986). It involves several stages: search a cloning of a gene of interest, sequencing, sequence modification (if needed, e.g. when codon usage in the gene is different from that in the model organism), gene transfer into the model organism, and studies of biochemical and phenotypic changes that entail expression of the foreign gene. Such an approach is feasible owing to the similarity of metabolic pathways and gene networks that regulate the activities of pro- and eukaryotic organisms under normal conditions and under exposure to various biotic and abiotic stresses. In addition, the use of bacterial genes helps to avoid many problems that arise during cloning, modifications and expression of eukaryotic genes in plants, whereas the constitutive nature of bacterial gene expression allows revealing "hot spots" of action of the homologous plant enzymes.

3. Usage of the methods of functional genomics for studying fundamental and applied aspects of plant life

3.1 Biotic stress tolerance

Activation tagging has been used for the isolation of mutants with resistance to biotic stress. For example, *CDR1-D* is a mutant that is resistant to sprayed suspensions of virulent *Pseudomonas syringae pathovar tomato* (*Pst*) (Xia et al. 2004). *CDR1* encodes an extracellular aspartic protease, which is a member of a large family of aspartic proteases in *Arabidopsis*. CDR1 functions in the production of a systemic signal that induces basal defenses. Another mutant, *FMO1-3D*, showed enhanced resistance to virulent *Pst* DC3000 (Koch et al. 2006). This phenotype is the result of the overexpression of a gene that encodes a class 3 FMO protein.

Recently we have proposed the model for studying the role of plant dioxygenases. Phenolic compounds serve as antioxidants and protect plants from active oxygen species. The content of phenolic compounds changes as plants grow and get mature and in response to biotic and abiotic influences, and these changes are achieved through modulation of enzymatic activities involved in their synthesis and degradation. Enzymes that take part in oxidation of aromatic compounds include dioxygenases (Tsoi et al. 1988). These enzymes oxidize phenolic compounds by breaking the aromatic ring, and thus enable subsequent biodegradation of phenols. There is evidence that plant dioxygenase (coded for by the *lls* gene of maize) may participate in the hypersensitive response of the plant to a pathogen attack (Lawton and Maleck 1998). For a study of the role played by dioxygenase in plants we have chosen the bacterial gene *nah*C (Y14173) of *Ps. putida*, coding for 1,2-dehydronaphtalene dioxygenase. Our choice was due to the fact that this enzyme possesses broad substrate specificity and can also use pyrocatechin as substrate (Tsoi et al. 1988), thus allowing to model a maximum number of dioxygenase isozymes. The expression of bacterial 1,2-dehydronaphthalene dioxygenase (coded by the *nah*C gene) in tobacco plants resulted in marked phenotypic and morphologic changes: chlorosis of the leaves, development of necrotic spots, delayed rooting and growth, and early flowering (Piruzian et al. 2002). Data on expression of bacterial 1,2-dihydroxynaphtalene dioxygenases in plants have not been reported in the literature. The necrotic spots on leaves of transgenic plants could have resulted from accumulation of phenolic substances. The above-mentioned phenotype and morphology changes suggested that the expression of bacterial dioxygenase resulted in alteration of the level of phenolic compounds in the transgenic plant cells. Measurements of phenolic acid content indicate that normal metabolism of phenolic compounds is disturbed in the plants, and the disturbance apparently results in induction of a stress response and appearance of the necrotic spots. In our opinion, such transgenic plants are a promising model for the study of mechanism of genome functioning under normal conditions and under stress, as well for the study of functions of phenolic compounds.

3.2 Abiotic stress tolerance

Environmental stresses are the major factors adversely affecting plant growth and development as well as productivity. Of the various abiotic stresses, drought and osmotic stress cause considerable agronomic problems by limiting crop yield and distribution world-wide (Chaves and Oliveira 2004).

Drought and osmotic stress induce a range of alterations at the molecular, biochemical, and cellular levels in plants, including stomatal closure, repression of photosynthesis, accumulation of osmolytes, and the inducible expression of genes involved in stress tolerance (Shinozaki and Yamaguchi-Shinozaki 2007).

The accumulation of proline by plants is a common physiological indicator and occurs under various abiotic stresses. There is an increasing body of evidence supporting the role of proline as a compatible osmolyte that maintains cellular osmotic adjustment and stabilizes the structure of proteins and membrane integrity (Verbruggen and Hermans 2008). Overexpression of different genes has been shown to significantly enhance proline levels in transgenic rice and improve their tolerance to environmental stresses (Ito et al. 2006; Liu et al. 2007; Pasquali et al. 2008; Xiang et al. 2007; Xu et al. 2008; Chen et al. 2009).

The transference of a single gene encoding a specific stress protein does not always result in sufficient expression to produce useful tolerance, because multiple and complex pathways are involved in controlling plant drought responses (Bohnert et al. 1995) and because modification of a single enzyme in a biochemical pathway is usually contrasted by a tendency of plant cells to restore homeostasis (Djilianov et al. 2002). Targeting multiple steps in a pathway may often modify metabolite fluxes in a more predictable manner. Another promising approach is therefore to engineer the overexpression of genes encoding stress inducible transcription factors.

There is increasingly more experimental support for the manipulation of the expression of stress-related transcription factor genes as a powerful tool in the engineering of stress-tolerant transgenic crops. This would, in turn, lead to the up-regulation of a series of stress-related genes under their control in transgenic plants (P. K. Agarwal et al. 2006).For example, the overexpression of transcription factor genes, such as ZFP252, SNAC1, OsNAC6, OsDREB1A, and HvCBF4, could enhance rice tolerance to different environmental stresses (Nakashima et al. 2007; Oh et al. 2007; Xiong et al. 2006; Xu et al. 2008; Yamaguchi-Shinozaki et al. 2006).

Following the application of microarray technology, several hundred stress induced genes, mainly in the model plant *Arabidopsis thaliana*, have been identified as candidates for manipulation (Shinozaki and Yamaguchi-Shinozaki 2007) and have been classified into three groups (Bhatnagar-Mathur et al. 2008): (a) genes encoding proteins with a known enzymatic or structural function. Examples include enzymes for synthesis of osmoprotective compounds, late embryogenesis abundant (LEA) proteins, osmotins, chaperons, channels involved in water movements through cell membranes, ubiquitins, proteases involved in protein turnover, and detoxifying enzymes; (b) genes with as yet unknown functions; and (c) regulatory genes, such as those coding for kinases, phosphatases and transcription factors.

Mutants with abiotic stress tolerance have been isolated by activation tagging and include the *edt1* mutant recently identified under drought conditions (Ahad et al. 2003; Ahad and Nick 2007; Pereira et al. 2004; Yu et al. 2008). This mutant showed a drought tolerant phenotype and reduced stomatal density. The enhanced drought tolerance of *edt1* was associated with an increase in the expression of the gene that encodes the transcription factor *HDG11*. The overexpression of *ArabidopsisHDG11* in tobacco can also confer drought tolerance and reduced leaf stomatal density (Zhang 2003).

FOX lines that consist of 43 stress-inducible transcription factors were constructed to elucidate stress-related gene function (Fujita et al. 2007). The T1 generation was screened for salt-stress-resistant lines and led to the identification of salt-tolerant lines. Among them, four lines harbored the same transgene, *AtbZIP60*, which encodes a basic domain/leucine zipper class transcription factor. The overexpression of *AtZIP60* leads to the upregulation of stress related genes, which suggests an important role for this transcription factor in stress-responsive signal transduction.

Transcription factors play an important role in plant development and stress responses. The *Arabidopsis* genome encodes more than 1,500 transcription factors. gain-of-function mutagenesis is an ideal approach to uncover the function of transcription factors (J. Z. Zhang 2003).

Weiste and colleagues (Weiste et al. 2007) generated an ORF collection composed of members of the ERF transcription factor family. They constructed a destination vector to enable ectopic expression driven by the*CaMV35S* promoter and included a HA-tag sequence to reveal transgene-specific expression. Using this library, they generated transgenic *Arabidopsis* plants that overexpress HA-tagged ORFs of the ERF transcription factor family. This approach yielded eight plants that show enhanced tolerance to oxidative stress resulting from the overexpression of the same ERF.

Typically a gene coding for a transcription factor in *Arabidopsis* is isolated, characterized and shown to improve drought response when overexpressed. The gene is then transferred to a crop plant where it often confers the same drought-tolerant phenotype. The HARDY (*HRD)* gene, coding for an AP2/ERF-like transcription factor (Pereira et al. 2007) is an example of this approach. *Arabidopsis* plants with a gain-of-function mutation in the *HRD* gene (hrd-D mutants) are drought resistant, salt-tolerant, and overexpress abiotic stress marker genes. Overexpression of the same gene in rice significantly improves water use efficiency both under well-watered conditions (50–100% increase) and under drought (50% increase). These plants also show enhanced photosynthetic assimilation and reduced transpiration (Pereira et al. 2007). *HRD* gene overexpression conserves drought tolerance in both dicots and monocots.

In other cases a gene coding for a transcription factor is isolated and characterized in *Arabidopsis*, but its orthologue gene in the crop plant of interest is identified and made to overexpress. For example Nelson et al. (Nelson et al. 2007) showed that overexpression of the *Arabidopsis* CAAT box-binding transcription factor AtNF-YB1 confers improved performance in *Arabidopsis* under drought conditions. They next overexpressed the orthologue of AtNF-YB1 (called ZmNF-YB2) in maize and found that, under simulated drought conditions, the altered maize plants produced up to 50% more than unmodified plants (Nelson et al. 2007).

A high-throughput gain-of-function approach has been applied to isolate salt stress tolerance genes using cDNAs of *Thellungiella halophila* (Du et al. 2008). *Thellungiella halophila* is a type of salt cress similar to *Arabidopsis* that can grow under high salt conditions. The cDNA library was prepared after salt stress treatment. Approximately 125,000 transgenic *Arabidopsis* that express *Thellungiella halophila* cDNAs under *CaMV 35S* promoter were generated. Novel salt stress tolerance genes were isolated from this mutant collection.

Ethylene response factor (ERF) genes have been successfully introduced into rice, generating transgenic rice with enhanced tolerance to biotic and abiotic stresses. For example, the tobacco OPBP1 (an AP2/ERF transcription factor) can enhance salt tolerance and disease resistance of transgenic rice (Chen and Guo 2008); ectopic expression of the *Arabidopsis* *HARDY* gene in rice improves water use efficiency and the ratio of biomass (Pereira et al. 2007); overexpression of rice DREB transcription factor (*OsDREB1F*) increases salt, drought, and low temperature tolerance in rice (Chu et al. 2008). Overexpression of transcription factor *Sub1A-1* in a submergence-intolerant *Oryza sativa* ssp. japonica conferred transgenic plants with enhanced submergence tolerance (Ronald et al. 2006). The ethylene response factors SNORKEL1 and SNORKEL2 encoding ERFs trigger internode elongation and allow rice to adapt to deep water (Ashikari et al. 2009).

By overexpressing a Athsp101 protein, Katiyar-Agarwal and associates (2003) generated a heat-tolerant transgenic rice (cv. Pusa basmati 1) line. This group showed that almost all the transgenic plants recovered after severe heat stress of 45–50⁰C and exhibited vigorous growth during the subsequent recovery at 28⁰C, while the untransformed plants could not recover to a similar extent.

In our experiments with salt stress tolerance, we have selected a mutant of *E. coli* able to grow on a medium with a high salt content. The cells of the mutant strain have a high content of proline, and it has been shown that this was due to a mutation in N-terminal region of γ-glutamyl kinase encoded by the *proB* gene. The mutation, which consists of single amino acid substitution (leucine is replaced by glycine) caused a conformational change in the regulatory region of the protein, made the enzyme less sensitive to the feedback inhibition by proline. We designated the gene coding for the mutant γ-glutamyl kinase as proB$_{osm}$ (Neumyvakin et al. 1990, 1991). *E. coli* genes *proB*$_{osm}$ and *proA* have been transferred into tobacco plants, each under control of a strong constitutive promoter CaMV 35S, which contains a duplicated enhancer sequence, or the P$_{mas}$ promoter which induces gene expression predominantly in roots (Sokhansandzh et al. 1997). The rationale for using a mutant form of the prokaryotic proteins which determine a strictly define phenotype, osmotolerance, was that the phenotype of the plant model will be easy to assay. The expression of the osmotolerance phenotype in the model eukaryotic organism would be indicative that the prokaryotic protein is an ortholog of the eukaryotic protein. Transgenic plants carrying the bacterial proline operon genes had a higher resistance to the toxic proline analogue (L-azetidin-2-carboxic acid) and to the high salt stress (were capable of rooting at NaCl concentrations in the medium over 350 mM) (Sokhansandzh et al. 1997). Thus, our results demonstrate usefulness of the proposed model and the possibility of simulating the activity of a bi-functional plant enzyme with two bacterial enzymes.

In *Arabidopsis*, knockout or silencing of *HSP101* caused loss of the acquired thermotolerance, whereas the overexpression of *HSP101* in transgenic plants improved tolerance to high temperature stress (Gurley 2000; Hong and Vierling 2000). Agarwal and co-workers (2003) provided evidence that *AtHSP101* and *OsHSP101* impart thermoprotection to yeast cells by dissolution of heat-induced protein aggregates. High-temperature-tolerant rice plants have also been produced by overexpressing a rice small heat-shock protein sHSP17.7 (Sato et al. 2004). Oxidative stress may accompany heat stress by the formation of ROS (Foolad et al. 2007). More recently, Qi and associates (2011) have reported that *mtHsp70* over-expression suppresses programmed cell death (PCD) by maintaining mitochondrial membrane potential and preventing ROS signal amplification in rice protoplasts.

Koh and co-workers (2007) reported that knockout (KO) mutants of rice *OsGSK1*, an orthologue of *Arabidopsis* BIN2, showed enhanced tolerance to several abiotic stresses including high temperature. In comparison to non-transgenic plants, the wilting ratios for knock out mutants were as much as 26% lower after heat (45^0C) stress. Feng and associates (2007) raised transgenic rice plants overexpressing rice sedoheptulose-1,7-bisphosphatase *SBPase*. They showed that overexpression of *SBPase* resulted in enhanced tolerance of growth and photosynthesis to high temperatures in transgenic rice plants.

Huang and co-workers (2008b) generated transgenic tobacco expressing rice A20/AN1- type zinc finger protein gene (*ZFP177*). Compared to wild type tobacco, the transgenic seedlings showed higher tolerance to temperature stress.

Major efforts have been made to identify genes that are associated with drought stress in a number of plant species (Gong et al. 2010; Huang et al. 2008a; Manavalan et al. 2009; Tran and Mochida 2010; Zheng et al. 2010). In rice, identification of drought-responsive genes has been carried out by means of expression profiling studies such as microarrays, expressed sequence tags (ESTs), RNA gel blot analyses and qRT-PCR (Rabbani et al. 2003; Rabello et al. 2008; Ramachandran et al. 2008; Reddy et al. 2007; Zhou et al. 2007). As a result, hundreds of genes that were induced or suppressed by drought stress have been identified. A number of these genes have been analyzed in detail, resulting in their characters as regulatory genes, such as transcription factor (TF) and protein kinase encoding genes, whose products regulate other stress-responsive genes. Some of the identified stress-responsive genes are functional genes which encode metabolic components, such as late embryogenesis abundant (LEA) proteins and osmoprotectant-synthesizing enzymes, important for stress tolerance (Yang et al. 2010).

Recently, Yang and associates (2010) classified drought-responsive genes into three groups based on their biological functions: transcriptional regulation, post-transcriptional RNA or protein phosphorylation, and osmoprotectant metabolism or molecular chaperons. However, among the genes that are affected by drought many genes have unknown functions. Efforts will be continued to determine the functions of the unknown drought-responsive genes.

Aquaporins, which are water channel proteins that translocate water across cell membranes, have been demonstrated for their roles in various physiological processes including stomatal closure (Li et al. 2008).

The rice plasma membrane intrinsic proteins (OsPIP) proteins are subfamilies of aquaporins and are divided into two subgroups, OsPIP1 and OsPIP2. Several members of OsPIP1 and OsPIP2 subfamilies were responsive to drought and salt stresses. Transgenic *Arabidopsis* overexpressing OsPIP2-2 showed enhanced tolerance to salt and drought stresses (Guo et al. 2006).

Transgenic rice overexpressing *Datura stramonium* S-adenosylmethionine decarboxylase (adc) gene showed increased tolerance to drought due to an increase in polyamine content (Capell et al. 2004), suggesting that OsAdc1 may be a potential candidate for development of enhanced drought-tolerant rice cultivars.

In some cases however, constitutive expression of a gene normally only induced by stress, has negative effects – so-called pleiotropic effects (Chan et al. 2002; Kasuga et al. 1999; Nakashima et al. 2007) – on growth and development when stress is not present. One solution is to use inducible (rather than constitutive) promoters that allow expression of a transgene only when it is required, while it is silenced otherwise. For example constitutive expression in *Arabidopsis* of *DREB1/CBF3*, a gene coding for a transcription factor induced by osmotic stress, confers tolerance to stress, but causes severe growth retardation under normal growth conditions (Kasuga et al. 1999). However, if this gene is expressed under the control of an osmotic stress-inducible promoter like RD29A no growth retardation occurs, and the plant is highly resistant to several stress conditions (Kasuga et al. 1999).Similarly,In tomato, overexpression of the *Arabidopsis CBF1* gene, encoding a transcription factor belonging to AP2/ERF family, confers increased drought, cold and oxidative stress tolerance compared to wild-type plants, but plant growth is severely affected (Chan et al. 2002). By contrast, when the same gene was placed under the control of a synthetic promoter derived from the barley *HVA22* gene, it was expressed mainly under abiotic stresses, so that the plant had the same tolerance characteristics towards stresses, but plant growth under normal conditions was not affected (Lee et al. 2003).

An ideal stress-inducible promoter would be completely silenced under normal conditions, but induced by stress in a fairly short time (a few hours) after stress onset. The promoter of the *Arabidopsis AtMYB41* gene, which is not expressed in any tissues under standard growth conditions but is highly induced in response to drought, salt and abscisic acid (Cominelli et al. 2008), may therefore be a very useful promoter.

3.3 Increase of productivity

The development of unique transgenic plants provides an applied angle, in making available highly nutritious "speciality crops," and also adds to the genetic resources that can be used to develop insightful knowledge base about genetic, biochemical, and physiological regulation of various metabolic pathways and functional metabolites. The transgenic tomatoes that accumulate higher polyamines, Spd and Spm, during ripening are a kind of a "gain of function" genotype, and we are using them to address the questions on the role of polyamines in fruit metabolism, in particular, their crosstalks with other functional molecules to enable higher nutritional quality of vegetables and fruits.

In this case, a fruit ripening-specific promoter was used to drive the expression of yeast *SAM decarboxylase* gene, with the result that the introduced gene was not active during the early growth and development of the plant but became active along with the normal ripening process of the fruit (Mehta et al. 2002).

An analysis of the principal, soluble constituents of wild-type and Spd/Spm-accumulating transgenic tomato, generated using high-resolution NMR spectroscopic methods, showed that the same metabolites were present in wild-type/azygous control tomatoes as in the transgenic tomato fruit. However, the latter conspicuously revealed differential metabolite content as compared to the controls (Mattoo et al. 2006). The red transgenic fruit were characterized by higher accumulation of the amino acids glutamine and asparagine; micronutrient choline; the organic acids citrate, fumarate, and malate; and an unidentified compound A. Compared to the control, wild-type fruit, the levels of valine, aspartic acid, sucrose, and glucose in the transgenic red fruit were reduced. These changes reflected specific alteration of metabolism, since the levels of isoleucine, glutamic acid, aminobutyric acid, phenylalanine, and fructose remained similar in the nontransgenic and transgenic fruits. Consequently, the transgenic red fruit have significantly higher fructose/glucose and acid [citrate+malate]/sugar [glucose+fructose+sucrose] ratios (Mattoo et al. 2006), consistent with higher fruit juice and nutritional quality reported in the 2 transgenics (Mehta et al. 2002), attributes favorably considered as higher quality in tomato breeding programs.

3.4 Xenobiotic tolerance

We have created transgenic plants expressing a mutant, glyphosate-resistant EPSP synthase from *E. coli* (Piruzian et al. 1988). The rationale for using a mutant form of the prokaryotic ortholog protein which determines a strictly define phenotype, herbicide resistance, was that the phenotype of the plant model will be easy to assay. The expression of the glyphosate tolerance phenotype in the model eukaryotic organism would be indicative that the prokaryotic protein is an ortholog of the eukaryotic protein. First we isolated an *E. coli* strain with resistance to glyphosate. The mutation was localized in locus aroA and was shown to result in a replacement of Ala with Pro in the bacterial EPSP synthase (Piruzian et al. 1988). We constructed expression vectors for plant transformation and obtained tobacco plants that tolerated a five-fold higher glyphosate concentration than inhibited the growth of control plants (Mett et al. 1991; Piruzian et al. 2000). Thus it was shown that with the use of transgenic plants expressing the mutant EPSP synthase, a bacterial enzyme insensitive to glyphosate could be active in the plant instead of the plant enzyme which was sensitive to the herbicide (Mett et al. 1991). This experiment also demonstrated a principal possibility of modeling plant enzymatic activity by using bacterial enzymes.

3.5 Studying processes of plant physiology

One of the best examples of the use of the activation-tagging system to identify genes involved in plant development is the isolation of *YUCCA* genes, which encode flavin mono-oxygenase (FMO) proteins involved in auxin biosynthesis. Six loci, which encode proteins with a role in auxin biosynthesis, have been identified in *Arabidopsis* by using activation-tagging technology (Pereira et al. 2002; Zhao et al. 2001). The *YUCCA* family consists of 11 members in the *Arabidopsis* genome (Cheng et al. 2006). All activation-tagged mutants of *YUCCA* genes showed phenotypes that were characteristic of auxin-overproducing mutants

(Boerjan et al. 1995; Delarue et al. 1998). Double, triple, and quadruple loss-of-function mutants of *YUCCA* genes showed deleterious developmental disorders, although no single mutants displayed visible phenotypes (Cheng et al. 2006). This functional redundancy might lead to difficulties in the isolation of mutants related to auxin biosynthesis by the loss-of-function approach. Thus, gain-of-function mutagenesis is a powerful tool to elucidate the function of genes that compose a gene family.

The *pap1-D Arabidopsis* mutant generated by activation tagging is an intense purple color caused by the overproduction of phenylpropanoid derivatives, such as anthocyanins (Borevitz et al. 2000). The *PAP1* gene encodes a member of the R2, R3 MYB transcription factor family that comprises more than 100 members in *Arabidopsis* (Paz-Ares et al. 1998; Weisshaar et al. 1998). The heterologous expression of *Arabidopsis PAP1* can enhance the accumulation of anthocyanins in tobacco plants. The activation-tagging method has also been employed to identify a transcriptional regulator of secondary metabolites in tomato (Mathews et al. 2003). The overexpression of *ANT1*, which encodes a MYB-type transcription factor, caused an intense purple color in many vegetative tissues throughout development and purple spotted fruit on the epidermis and pericarp in tomato.

LeClere and Bartel (2001) generated *Arabidopsis* lines that overexpresses random cDNAs driven by the *CaMV 35S* promoter. They generated more than 30,000 *Arabidopsis* transgenic plants and isolated a mutant that showed a pale green phenotype caused by the overexpression of a truncated cDNA that encodes chloroplast ferredoxin-NADP+ reductase (FNR). This phenotype was caused by the cosuppression of endogenous genes by transgene overexpression, which led to dominant loss-of-function phenotypes.

Using *Arabidopsis* FOX lines, Okazaki and colleagues isolated six mutants that contained an increased number of chloroplasts in their leaves (2009). The overproduction of plastid division (PDV) proteins that regulate the rate of chloroplast division in *Arabidopsis* leads to an increase in the number, but a decrease in the size, of chloroplasts.

Another *Arabidopsis* FOX line that carries the cDNA, which encodes the cytokinin responsive transcription factor 2 (CRF2), also showed an increased number of chloroplasts.

Thus, the FOX hunting system is capable of the highthroughput characterization of gene functions.

A high-efficiency transformation method has been developed in rice. This makes rice an ideal host plant for the FOX hunting system (Ichikawa et al. 2007). More than 28,000 rice fl-cDNAs have been generated (Kikuchi et al. 2003). Approximately 12,000 rice lines have been generated in which 13,980 independent fl-cDNAs were overexpressed under the control of the ubiquitin promoter. Among several phenotypes in the T0 generation, three dwarf lines that carry the same novel *gibberellin 2-oxidase* (*GA2ox*) gene were isolated.

The ORF overexpression approach can also be used to investigate the function of putative genes identified by computer-based means. Small secreted peptides, less than 150 amino acids long, were predicted to identify genes involved in plant development in *Arabidopsis* (Hara et al. 2007). Plants that overexpress 153 predicted genes that encode these small secreted peptides were generated, and this approach led to the identification of the *Epidermal Patterning Factor 1* (*EPF1*) gene. *EPF1* was expressed in stomatal cells, and precursors may be involved in the control of stomatal patterning through the regulation of asymmetric cell division.

The T-DNA vector pER16, which contains the estradiol-inducible promoter, was used for the conditional activation of nearby genes by the addition of estradiol. This system was used to identify the gene *PGA6*, which encodes WUSCHEL (WUS), a homeodomain protein involved in the regulation of stem cell fate in *Arabidopsis* shoot and floral meristems (Mayer et al. 1998).

Recently we have proposed a new strategy for creating experimental models for plant functional genomics. It is based on the expression in transgenic plants of genes from thermophilic bacteria encoding functional analogues of plant proteins with high specific activity and thermal stability. We have validated this strategy by comparing physiological, biochemical and molecular properties of control tobacco plants and transgenic plants expressing genes of β-glucanases with different substrate specificity. We demonstrate that the expression of bacterial β-1,3–1,4-glucanase gene exerts no significant influence on tobacco plant metabolism, while the expression of bacterial β-1,3-glucanase affects plant metabolism only at early stages of growth and development. By contrast, the expression of bacterial β-1,4-glucanase has a significant effect on transgenic tobacco plant metabolism, namely, it affects plant morphology, the thickness of the primary cell wall, phytohormonal status, and the relative sugar content. We propose a hypothesis of β-glucanase action as an important factor of genetic regulation of metabolic processes in plants.

It should also be mentioned that many plant enzymes have numerous isozymes, and for this reason the activity assays as well as functional studies of particular isozymes *in vivo* are difficult (del Campillo 1999; Libertini et al. 2004). It should be noted that it was the use of thermostable heterologous proteins that enabled us to obtain these results. Overexpression of a homologous plant gene or a gene from a related species, apart from the additional difficulties of cloning plant genes with their exon–intron structure, could result in technical problems in detecting and assessing the activity of these proteins in transgenic plants. Thus, the use of thermostable bacterial proteins–functional analogs of plant beta-glucanase is not only adequate but also a more convenient method when compared to over-expression of a plant gene from its own genome or that of a related species.

The next our work was the construction of experimental models for studying the role of isopentyl transferases in phytohormone synthesis and plant differentiation.

It has been supposed that phytohormones, cytokinins in particular, are largely responsible for the viability of plants following exposure to abiotic stress and to pathogens. Therefore, an employment of genes whose expression alters the phytohormone balance for studying plant metabolism is deemed especially promising. One of such enzymes is isopentenyl transferase, a key enzyme of the cytokinin biosynthesis pathway. As a functional bacterial analogue of this enzyme, we have used isopentyl transferase (coded for by the T-*cyt* or *ipt* gene), the key enzyme of cytokinin synthesis from the T region of the *Agrobacterium tumefaciens* Ti plasmid. The product of the *ipt* gene is involved in crown gall formation in plants. In accordance with our strategy, we cloned the *ipt* gene and introduced it into the tobacco genome (Iusibov et al. 1989). Regenerated shoots at first did not form roots and were devoid of apical domination, a fact that was also reported by others. As reported by Zhang and co-workers, (1996), the expression of isopentyl transferase in transgenic tobacco plants may be controlled by auxins. We have therefore used exogenous auxin and this allowed us to obtain normal transgenic plants having an intermediate level of cytokinin in comparison with normal transgenic plants and the crown gall tissue (Makarova et al. 1997).

On the whole, expression of the agrobacterial gene leads to cytokinin overproduction (a two-fold excess of total cytokinins), decrease of the abscisic acid level, elevated level of chlorogenic acid, and disturbed morphogenesis and regeneration processes in the plants (Makarova et al. 1997; Yusibov et al. 1991). The altered hormone balance naturally affected such a vitally important process as photosynthesis. In particular, we have shown that elevated cytokinin affects the expression of some plant genes. For example, we found that plants transgenic for the *ipt* gene had a higher level of mRNA of the chloroplast gene of the ribulose biphosphate carboxylase (RBC) smaller subunit (*rbcL*) (Yusibov et al. 1991). Thus, the expression of bacterial isopentyl transferase causes significant metabolic changes in the transgenic plants. These include altered hormone balance (cytokinins, abscisic acid), altered expression of some chloroplast genes (the RBC smaller subunit) involved in photosynthesis, and altered morphology of the plants.

To study the role of enzymes related to hydrocarbon metabolism in plants we have chosen the gene *xyl*A *of E. coli*. This gene codes for xylose (glucose) isomerase (P00944) (EC 5.3.1.5) which converts xylose into xylulose and *vice versa* as well as fructose into glucose and *vice versa*. *E. coli' s* enzyme is thermostable (Piruzian et al. 1989), a property which ensures an easy assay of the enzyme in plants. The *xyl*A gene under control of the 35S CaMV promoter was transferred to tobacco plants using the *A. tumefaciens* vector system, and it was shown that an active enzyme is produced in the transgenic plants (Goldenkova et al. 2002). The plants had larger leaves, grew faster and had stronger roots than the controls. The expression of bacterial xylose (glucose) isomerase induces morphological changes in transgenic plants that correlate with changes in the expression of chloroplast genes involved in photosynthesis and maintaining the phytohormone balance. Thus transgenic plants of the XylA type represent a promising model system for studying photosynthesis as a function of phytohormone activity.

4. Conclusion

In plant functional genomics most approaches have introduced genes with a constitutive or inducible promoters, resulting in gene overexpression in transgenic plants. In some cases, however, it has been conferred by gene down-regulation by RNA interference, co-suppression or loss-of-function mutants. Each approach has advantages and disadvantages in different aspects of high-throughput characterization of gene functions. The first aspect is the basic construction strategy to produce a large population of mutant lines. The second aspect is whether mutants generated in each system can cover the vast numbers and wide variety of genes. Identification of all gene functions is the final goal of functional analysis at a genome level, and the production of mutants for this purpose. The final aspect is the enhancement of endogenous gene expression with tissue specificity. Highthroughput functional genomics is helping to shift the focus from the characterization of individual gene functions to a more systems-based holistic or synthetic approach to understand the genetic mechanisms that underlay gene regulation and complex signaling networks.

5. Acknowledgment

This work was supported by grant from Russian Academy of Sciences ("Biodiversity program"), and grant Russian Foundation for Basic Research (grant # 10-04-01195-Б).

6. References

Agarwal, M., et al. (2003), 'Molecular characterization of rice hsp101: complementation of yeast hsp104 mutation by disaggregation of protein granules and differential expression in indica and japonica rice types', *Plant Mol Biol*, 51 (4), 543-53.

Agarwal, P. K., et al. (2006), 'Role of DREB transcription factors in abiotic and biotic stress tolerance in plants', *Plant Cell Reports*, 25 (12), 1263-74.

Ahad, A. and Nick, P. (2007), 'Actin is bundled in activation-tagged tobacco mutants that tolerate aluminum', *Planta*, 225 (2), 451-68.

Ahad, A., Wolf, J., and Nick, P. (2003), 'Activation-tagged tobacco mutants that are tolerant to antimicrotubular herbicides are cross-resistant to chilling stress', *Transgenic Research*, 12 (5), 615-29.

Anonymous (2000), 'Analysis of the genome sequence of the flowering plant Arabidopsis thaliana', *Nature*, 408 (6814), 796-815.

Ashikari, M., et al. (2009), 'The ethylene response factors SNORKEL1 and SNORKEL2 allow rice to adapt to deep water', *Nature*, 460 (7258), 1026-U116.

Bechtold, N. and Pelletier, G. (1998), 'In planta Agrobacterium-mediated transformation of adult Arabidopsis thaliana plants by vacuum infiltration', *Methods Mol Biol*, 82, 259-66.

Bhatnagar-Mathur, P., Vadez, V., and Sharma, K. K. (2008), 'Transgenic approaches for abiotic stress tolerance in plants: retrospect and prospects', *Plant Cell Rep*, 27 (3), 411-24.

Birnbaum, K., et al. (2003), 'A gene expression map of the Arabidopsis root', *Science*, 302 (5652), 1956-60.

Birnbaum, K., et al. (2005), 'Cell type-specific expression profiling in plants via cell sorting of protoplasts from fluorescent reporter lines', *Nat Methods*, 2 (8), 615-9.

Boerjan, W., et al. (1995), 'Superroot, a recessive mutation in Arabidopsis, confers auxin overproduction', *Plant Cell*, 7 (9), 1405-19.

Bohnert, H. J., Nelson, D. E., and Jensen, R. G. (1995), 'Adaptations to Environmental Stresses', *Plant Cell*, 7 (7), 1099-111.

Borevitz, J. O., et al. (2000), 'Activation tagging identifies a conserved MYB regulator of phenylpropanoid biosynthesis', *Plant Cell*, 12 (12), 2383-94.

Brady, S. M., et al. (2007), 'A high-resolution root spatiotemporal map reveals dominant expression patterns', *Science*, 318 (5851), 801-6.

Brand, L., et al. (2006), 'A versatile and reliable two-component system for tissue-specific gene induction in Arabidopsis', *Plant Physiol*, 141 (4), 1194-204.

Capell, T., Bassie, L., and Christou, P. (2004), 'Modulation of the polyamine biosynthetic pathway in transgenic rice confers tolerance to drought stress', *Proc Natl Acad Sci U S A*, 101 (26), 9909-14.

Carninci, P., et al. (1998), 'Thermostabilization and thermoactivation of thermolabile enzymes by trehalose and its application for the synthesis of full length cDNA', *Proc Natl Acad Sci U S A*, 95 (2), 520-4.

Carninci, P., et al. (1997), 'High efficiency selection of full-length cDNA by improved biotinylated cap trapper', *DNA Res*, 4 (1), 61-6.

Carninci, P., et al. (1996), 'High-efficiency full-length cDNA cloning by biotinylated CAP trapper', *Genomics*, 37 (3), 327-36.

Chan, M. T., et al. (2002), 'Tomato plants ectopically expressing Arabidopsis CBF1 show enhanced resistance to water deficit stress', *Plant Physiology*, 130 (2), 618-26.

Chan, S. W. L. (2008), 'Inputs and outputs for chromatin-targeted RNAi', *Trends in Plant Science*, 13 (7), 383-89.

Chaves, M. M. and Oliveira, M. M. (2004), 'Mechanisms underlying plant resilience to water deficits: prospects for water-saving agriculture', *J Exp Bot*, 55 (407), 2365-84.

Chen, J. B., et al. (2009), 'Cloning the PvP5CS gene from common bean (Phaseolus vulgaris) and its expression patterns under abiotic stresses', *J Plant Physiol*, 166 (1), 12-9.

Chen, X. and Guo, Z. (2008), 'Tobacco OPBP1 enhances salt tolerance and disease resistance of transgenic rice', *Int J Mol Sci*, 9 (12), 2601-13.

Cheng, Y., Dai, X., and Zhao, Y. (2006), 'Auxin biosynthesis by the YUCCA flavin monooxygenases controls the formation of floral organs and vascular tissues in Arabidopsis', *Genes Dev*, 20 (13), 1790-9.

Chu, C. C., et al. (2008), 'Overexpression of a rice OsDREB1F gene increases salt, drought, and low temperature tolerance in both Arabidopsis and rice', *Plant Molecular Biology*, 67 (6), 589-602.

Collins, J. E., et al. (2004), 'A genome annotation-driven approach to cloning the human ORFeome', *Genome Biol*, 5 (10), R84.

Cominelli, E., et al. (2008), 'Over-expression of the Arabidopsis AtMYB41 gene alters cell expansion and leaf surface permeability', *Plant Journal*, 53 (1), 53-64.

del Campillo, E. (1999), 'Multiple endo-1,4-beta-D-glucanase (cellulase) genes in Arabidopsis', *Curr Top Dev Biol*, 46, 39-61.

Delarue, M., et al. (1998), 'Sur2 mutations of Arabidopsis thaliana define a new locus involved in the control of auxin homeostasis', *Plant Journal*, 14 (5), 603-11.

Dinneny, J. R. (2008), 'Cell identity mediates the response of Arabidopsis roots to abiotic stress (vol 320, pg 942, 2008)', *Science*, 322 (5898), 44-44.

Dinneny, J. R. (2010), 'Analysis of the salt-stress response at cell-type resolution', *Plant Cell Environ*, 33 (4), 543-51.

Dinneny, J. R., et al. (2008), 'Cell identity mediates the response of Arabidopsis roots to abiotic stress', *Science*, 320 (5878), 942-5.

Djilianov, D., et al. (2002), 'Freezing tolerant tobacco, transformed to accumulate osmoprotectants', *Plant Science*, 163 (1), 157-64.

Du, J., et al. (2008), 'Functional gene-mining for salt-tolerance genes with the power of Arabidopsis', *Plant Journal*, 56 (4), 653-64.

Eamens, A., et al. (2008), 'RNA silencing in plants: Yesterday, today, and tomorrow', *Plant Physiology*, 147 (2), 456-68.

Feng, L., et al. (2007), 'Overexpression of SBPase enhances photosynthesis against high temperature stress in transgenic rice plants', *Plant Cell Rep*, 26 (9), 1635-46.

Fire, A., et al. (1998), 'Potent and specific genetic interference by double-stranded RNA in Caenorhabditis elegans', *Nature*, 391 (6669), 806-11.

Foolad, M. R., et al. (2007), 'Heat tolerance in plants: An overview', *Environmental and Experimental Botany*, 61 (3), 199-223.

Frommer, W. B., Davidson, M. W., and Campbell, R. E. (2009), 'Genetically encoded biosensors based on engineered fluorescent proteins', *Chem Soc Rev*, 38 (10), 2833-41.

Fujita, M., et al. (2007), 'Identification of stress-tolerance-related transcription-factor genes via mini-scale Full-length cDNA Over-eXpressor (FOX) gene hunting system', *Biochem Biophys Res Commun*, 364 (2), 250-7.

Fuller, C. W., et al. (2009), 'The challenges of sequencing by synthesis', *Nat Biotechnol*, 27 (11), 1013-23.

Gifford, M. L., et al. (2008), 'Cell-specific nitrogen responses mediate developmental plasticity', *Proc Natl Acad Sci U S A*, 105 (2), 803-8.

Goff, S. A., et al. (2002), 'A draft sequence of the rice genome (Oryza sativa L. ssp japonica)', *Science*, 296 (5565), 92-100.

Goldenkova, I.V., et al. (2002), 'The Expression of the Bacterial Gene for Xylose(Glucose) Isomerase in Transgenic Tobacco Plants Affects Plant Morphology and Phytohormonal Balance', *Russ. J. Plant Physiol.*, 49 (4), 524-29.

Gong, P., et al. (2010), 'Transcriptional profiles of drought-responsive genes in modulating transcription signal transduction, and biochemical pathways in tomato', *J Exp Bot*, 61 (13), 3563-75.

Guo, L., et al. (2006), 'Expression and functional analysis of the rice plasma-membrane intrinsic protein gene family', *Cell Res*, 16 (3), 277-86.

Gurley, W. B. (2000), 'HSP101: a key component for the acquisition of thermotolerance in plants', *Plant Cell*, 12 (4), 457-60.

Hamilton, A. J. and Baulcombe, D. C. (1999), 'A species of small antisense RNA in posttranscriptional gene silencing in plants', *Science*, 286 (5441), 950-2.

Hara, K., et al. (2007), 'The secretory peptide gene EPF1 enforces the stomatal one-cell-spacing rule', *Genes Dev*, 21 (14), 1720-5.

Heyman, J. A., et al. (1999), 'Genome-scale cloning and expression of individual open reading frames using topoisomerase I-mediated ligation', *Genome Research*, 9 (4), 383-92.

Hong, S. W. and Vierling, E. (2000), 'Mutants of Arabidopsis thaliana defective in the acquisition of tolerance to high temperature stress', *Proc Natl Acad Sci U S A*, 97 (8), 4392-7.

Huang, D., et al. (2008a), 'The relationship of drought-related gene expression in Arabidopsis thaliana to hormonal and environmental factors', *J Exp Bot*, 59 (11), 2991-3007.

Huang, J., et al. (2008b), 'Expression analysis of rice A20/AN1-type zinc finger genes and characterization of ZFP177 that contributes to temperature stress tolerance', *Gene*, 420 (2), 135-44.

Ichikawa, H., et al. (2007), 'A genome-wide gain-of-function analysis of rice genes using the FOX-hunting system', *Plant Molecular Biology*, 65 (4), 357-71.

Ito, T. and Meyerowitz, E. M. (2000), 'Overexpression of a gene encoding a cytochrome P450, CYP78A9, induces large and seedless fruit in arabidopsis', *Plant Cell*, 12 (9), 1541-50.

Ito, Y., et al. (2006), 'Functional analysis of rice DREB1/CBF-type transcription factors involved in cold-responsive gene expression in transgenic rice', *Plant Cell Physiol*, 47 (1), 141-53.

Iusibov, V. M., et al. (1989), '[Transfer of the agrobacterial gene for cytokinin biosynthesis into tobacco plants]', *Mol Gen Mikrobiol Virusol*, (7), 11-3.

Iyer-Pascuzzi, A., et al. (2009), 'Functional genomics of root growth and development in Arabidopsis', *Curr Opin Plant Biol*, 12 (2), 165-71.

Jaillon, O., et al. (2007), 'The grapevine genome sequence suggests ancestral hexaploidization in major angiosperm phyla', *Nature*, 449 (7161), 463-7.

Kakimoto, T. (1996), 'CKI1, a histidine kinase homolog implicated in cytokinin signal transduction', *Science*, 274 (5289), 982-5.

Kasuga, M., et al. (1999), 'Improving plant drought, salt, and freezing tolerance by gene transfer of a single stress-inducible transcription factor', *Nature Biotechnology*, 17 (3), 287-91.

Katiyar-Agarwal, S., Agarwal, M., and Grover, A. (2003), 'Heat-tolerant basmati rice engineered by over-expression of hsp101', *Plant Mol Biol*, 51 (5), 677-86.

Kaul, S., et al. (2000), 'Analysis of the genome sequence of the flowering plant Arabidopsis thaliana', *Nature*, 408 (6814), 796-815.

Kikuchi, S., et al. (2003), 'Collection, mapping, and annotation of over 28,000 cDNA clones from japonica rice', *Science*, 301 (5631), 376-9.

Kitagawa, M., et al. (2005), 'Complete set of ORF clones of Escherichia coli ASKA library (a complete set of E. coli K-12 ORF archive): unique resources for biological research', *DNA Res*, 12 (5), 291-9.

Koch, M., et al. (2006), 'A role for a flavin-containing mono-oxygenase in resistance against microbial pathogens in Arabidopsis', *Plant Journal*, 47 (4), 629-39.

Koh, J. H., et al. (2007), 'T-DNA tagged knockout mutation of rice OsGSK1, an orthologue of Arabidopsis BIN2, with enhanced tolerance to various abiotic stresses', *Plant Molecular Biology*, 65 (4), 453-66.

Lander, E. S., et al. (2001), 'Initial sequencing and analysis of the human genome', *Nature*, 409 (6822), 860-921.

Lawton, K. and Maleck, K. (1998), 'Plant Strategies for Resistance to Pathogens ', *Curr Opin Biotechnol*, 9, 208-13.

LeClere, S. and Bartel, B. (2001), 'A library of Arabidopsis 35S-cDNA lines for identifying novel mutants', *Plant Mol Biol*, 46 (6), 695-703.

Lee, J. T., et al. (2003), 'Expression of Arabidopsis CBF1 regulated by an ABA/stress inducible promoter in transgenic tomato confers stress tolerance without affecting yield', *Plant Cell and Environment*, 26 (7), 1181-90.

Li, G. W., et al. (2008), 'Characterization of OsPIP2;7, a water channel protein in rice', *Plant Cell Physiol*, 49 (12), 1851-8.

Libertini, E., Li, Y., and McQueen-Mason, S. J. (2004), 'Phylogenetic analysis of the plant endo-beta-1,4-glucanase gene family', *J Mol Evol*, 58 (5), 506-15.

Liu, K., et al. (2007), 'Overexpression of OsCOIN, a putative cold inducible zinc finger protein, increased tolerance to chilling, salt and drought, and enhanced proline level in rice', *Planta*, 226 (4), 1007-16.

Makarova, R.V., et al. (1997), 'Phytohormone Production in Tobacco ipt-Regenerates in vitro', *Russ. J. Plant Physiol.*, 44 (5), 762-68.

Manavalan, L. P., et al. (2009), 'Physiological and molecular approaches to improve drought resistance in soybean', *Plant Cell Physiol*, 50 (7), 1260-76.

Mathews, H., et al. (2003), 'Activation tagging in tomato identifies a transcriptional regulator of anthocyanin biosynthesis, modification, and transport', *Plant Cell*, 15 (8), 1689-703.

Matsui, M., et al. (2006), 'The FOX hunting system: an alternative gain-of-function gene hunting technique', *Plant Journal*, 48 (6), 974-85.

Matsui, M., et al. (2009), 'Systematic approaches to using the FOX hunting system to identify useful rice genes', *Plant Journal*, 57 (5), 883-94.

Matsuyama, A., et al. (2006), 'ORFeome cloning and global analysis of protein localization in the fission yeast Schizosaccharomyces pombe (vol 24, pg 841, 2006)', *Nature Biotechnology*, 24 (8), 1033-33.

Mattoo, A. K., et al. (2006), 'Nuclear magnetic resonance spectroscopy-based metabolite profiling of transgenic tomato fruit engineered to accumulate spermidine and spermine reveals enhanced anabolic and nitrogen-carbon interactions', *Plant Physiol*, 142 (4), 1759-70.

Mayer, K. F., et al. (1998), 'Role of WUSCHEL in regulating stem cell fate in the Arabidopsis shoot meristem', *Cell*, 95 (6), 805-15.

McGinnis, K., et al. (2005), 'Transgene-induced RNA interference as a tool for plant functional genomics', *Rna Interference*, 392, 1-24.

Mehta, R. A., et al. (2002), 'Engineered polyamine accumulation in tomato enhances phytonutrient content, juice quality, and vine life', *Nature Biotechnology*, 20 (6), 613-8.

Mello, C. C., et al. (2001), 'Genes and mechanisms related to RNA interference regulate expression of the small temporal RNAs that control C-elegans developmental timing', *Cell*, 106 (1), 23-34.

Mett, V.L., et al. (1991), 'Cloning and Expression of Mutant Gene for EPSP-Syntase Escherichia coli in Transgenic Plants.', *Rus. J. Biotechnology (Rus.)*, 3, 19-22.

Mette, M. F., et al. (2000), 'Transcriptional silencing and promoter methylation triggered by double-stranded RNA', *Embo Journal*, 19 (19), 5194-201.

Metzker, M. L. (2010), 'Sequencing technologies - the next generation', *Nat Rev Genet*, 11 (1), 31-46.

Ming, R., et al. (2008), 'The draft genome of the transgenic tropical fruit tree papaya (Carica papaya Linnaeus)', *Nature*, 452 (7190), 991-6.

Mittler, R., Feng, X., and Cohen, M. (1998), 'Post-transcriptional suppression of cytosolic ascorbate peroxidase expression during pathogen-induced programmed cell death in tobacco', *Plant Cell*, 10 (3), 461-73.

Nakashima, K., et al. (2007), 'Functional analysis of a NAC-type transcription factor OsNAC6 involved in abiotic and biotic stress-responsive gene expression in rice', *Plant Journal*, 51 (4), 617-30.

Nakazawa, M., et al. (2001), 'DFL1, an auxin-responsive GH3 gene homologue, negatively regulates shoot cell elongation and lateral root formation, and positively regulates the light response of hypocotyl length', *Plant Journal*, 25 (2), 213-21.

Nakazawa, M., et al. (2003), 'Activation tagging, a novel tool to dissect the functions of a gene family', *Plant Journal*, 34 (5), 741-50.

Nanjo, T., et al. (2007), 'Functional annotation of 19,841 Populus nigra full-length enriched cDNA clones', *BMC Genomics*, 8, 448.

Napoli, C., Lemieux, C., and Jorgensen, R. (1990), 'Introduction of a Chimeric Chalcone Synthase Gene into Petunia Results in Reversible Co-Suppression of Homologous Genes in Trans', *Plant Cell*, 2 (4), 279-89.

Nelson, D. E., et al. (2007), 'Plant nuclear factor Y (NF-Y) B subunits confer drought tolerance and lead to improved corn yields on water-limited acres', *Proceedings of the National Academy of Sciences of the United States of America*, 104 (42), 16450-55.

Neumyvakin, L.V., Kobets, N.S., and Piruzian, E.S. (1990), 'Cultivation of an Escherichia coli Mutant Superproducing Proline and Resistant to Increased Concentration of NaCl. ', *Rus. J. Genetics* 26 (8), 1370-79.

Neumyvakin, L.V., Kobets, N.S., and Piruzian, E.S. (1991), 'Obtaining and Cloning Mutant proBosm Gene from Escherichia coli Providing the Resistant to Increased Concentration of NaCl. ', *Rus. J. Cytology* 33, 122.

Nishiyama, T., et al. (2003), 'Comparative genomics of Physcomitrella patens gametophytic transcriptome and Arabidopsis thaliana: implication for land plant evolution', *Proc Natl Acad Sci U S A*, 100 (13), 8007-12.

Odell, J. T., Nagy, F., and Chua, N. H. (1985), 'Identification of DNA sequences required for activity of the cauliflower mosaic virus 35S promoter', *Nature*, 313 (6005), 810-2.

Ogihara, Y., et al. (2004), 'Construction of a full-length cDNA library from young spikelets of hexaploid wheat and its characterization by large-scale sequencing of expressed sequence tags', *Genes Genet Syst*, 79 (4), 227-32.

Oh, S. J., et al. (2007), 'Expression of barley HvCBF4 enhances tolerance to abiotic stress in transgenic rice', *Plant Biotechnol J*, 5 (5), 646-56.

Okazaki, K., et al. (2009), 'The PLASTID DIVISION1 and 2 components of the chloroplast division machinery determine the rate of chloroplast division in land plant cell differentiation', *Plant Cell*, 21 (6), 1769-80.

Ostergaard, L. and Yanofsky, M. F. (2004), 'Establishing gene function by mutagenesis in Arabidopsis thaliana', *Plant Journal*, 39 (5), 682-96.

Pasquali, G., et al. (2008), 'Osmyb4 expression improves adaptive responses to drought and cold stress in transgenic apples', *Plant Cell Rep*, 27 (10), 1677-86.

Paterson, A. H., et al. (2009), 'The Sorghum bicolor genome and the diversification of grasses', *Nature*, 457 (7229), 551-6.

Paz-Ares, J., et al. (1998), 'More than 80R2R3-MYB regulatory genes in the genome of Arabidopsis thaliana', *Plant Journal*, 14 (3), 273-84.

Pereira, A., et al. (2002), 'Activation tagging using the En-I maize transposon system in Arabidopsis', *Plant Physiology*, 129 (4), 1544-56.

Pereira, A., et al. (2004), 'The SHINE clade of AP2 domain transcription factors activates wax biosynthesis, alters cuticle properties, and confers drought tolerance when overexpressed in Arabidopsis', *Plant Cell*, 16 (9), 2463-80.

Pereira, A., et al. (2007), 'Improvement of water use efficiency in rice by expression of HARDY, an Arabidopsis drought and salt tolerance gene', *Proceedings of the National Academy of Sciences of the United States of America*, 104 (39), 15270-75.

Piano, F., et al. (2005), 'New genes with roles in the C-elegans embryo revealed using RNAi of ovary-enriched ORFeome clones', *Genome Research*, 15 (2), 250-59.

Piruzian, E. S. and Andrianov, V. M. (1986), 'Cloning and analysis of replication region of the *Agrobacterium tumefaciens* C58 nopaline Ti plasmid and its application for foreign gene transfer into plants', *Genetika (Russian)*, 22, 2674-83.

Piruzian, E. S., Stekhin, I. N., and Andrianov, V. M. (1983), '[Cloning of the Ti-plasmid DNA fragments of Agrobacterium tumefaciens in Escherichia coli and the identification of the site of origin of Ti-plasmid replication]', *Dokl Akad Nauk SSSR*, 273 (5), 1249-51.

Piruzian, E. S., et al. (1988), 'The use of bacterial genes encoding herbicide tolerance in constructing transgenic plants', *Microbiol Sci*, 5 (8), 242-8.

Piruzian, E. S., et al. (1989), '[Expression of the Escherichia coli glucose isomerase gene in transgenic plants]', *Dokl Akad Nauk SSSR*, 305 (3), 729-31.

Piruzian, E.S., et al. (2002), 'Physiological and Biochemical Characteristics of Tobacco Transgenic Plants Expressing Bacterial Dioxygenase ', *Russ. J. Plant Physiol.* , 49, 817-22.

Piruzian, E.S., et al. (2000), 'Transgenic Plants Expressing Foreign Genes as a Model for Studying Plant Stress Responses and a Source for Resistant Plant Forms', *Russian J. Plant Physiol.* , 47 (3), 327-36.

Qi, Y., et al. (2011), 'Over-expression of mitochondrial heat shock protein 70 suppresses programmed cell death in rice', *FEBS Lett*, 585 (1), 231-9.

Rabbani, M. A., et al. (2003), 'Monitoring expression profiles of rice genes under cold, drought, and high-salinity stresses and abscisic acid application using cDNA microarray and RNA gel-blot analyses', *Plant Physiol*, 133 (4), 1755-67.

Rabello, A. R., et al. (2008), 'Identification of drought-responsive genes in roots of upland rice (Oryza sativa L)', *BMC Genomics*, 9, 485.

Ralph, S. G., et al. (2008), 'A conifer genomics resource of 200,000 spruce (Picea spp.) ESTs and 6,464 high-quality, sequence-finished full-length cDNAs for Sitka spruce (Picea sitchensis)', *BMC Genomics*, 9, 484.

Ramachandran, S., et al. (2008), 'A comprehensive transcriptional profiling of the WRKY gene family in rice under various abiotic and phytohormone treatments', *Plant and Cell Physiology*, 49 (6), 865-79.

Reddy, A. R., et al. (2007), 'Identification of stress-responsive genes in an indica rice (Oryza sativa L.) using ESTs generated from drought-stressed seedlings', *Journal of Experimental Botany*, 58 (2), 253-65.

Ronald, P. C., et al. (2006), 'Sub1A is an ethylene-response-factor-like gene that confers submergence tolerance to rice', *Nature*, 442 (7103), 705-08.

Rual, J. F., et al. (2004), 'Human ORFeome version 1.1: a platform for reverse proteomics', *Genome Research*, 14 (10B), 2128-35.

Sakurai, T., et al. (2005), 'RARGE: a large-scale database of RIKEN Arabidopsis resources ranging from transcriptome to phenome', *Nucleic Acids Res*, 33 (Database issue), D647-50.

Sakurai, T., et al. (2007), 'Sequencing analysis of 20,000 full-length cDNA clones from cassava reveals lineage specific expansions in gene families related to stress response', *BMC Plant Biol*, 7, 66.

Sasaki, T., et al. (2002), 'The genome sequence and structure of rice chromosome 1', *Nature*, 420 (6913), 312-6.

Sato, K., et al. (2009), 'Development of 5006 full-length CDNAs in barley: a tool for accessing cereal genomics resources', *DNA Res*, 16 (2), 81-9.

Sato, Y., et al. (2004), 'Over-expression of a small heat shock protein, sHSP17.7, confers both heat tolerance and UV-B resistance to rice plants', *Molecular Breeding*, 13 (2), 165-75.

Seki, M., et al. (2002), 'Functional annotation of a full-length Arabidopsis cDNA collection', *Science*, 296 (5565), 141-5.

Shinozaki, K. and Yamaguchi-Shinozaki, K. (2007), 'Gene networks involved in drought stress response and tolerance', *J Exp Bot*, 58 (2), 221-7.

Singer, T. and Burke, E. (2003), 'High-throughput TAIL-PCR as a tool to identify DNA flanking insertions', *Methods Mol Biol*, 236, 241-72.

Slakeski, N., et al. (1990), 'Structure and tissue-specific regulation of genes encoding barley (1----3, 1----4)-beta-glucan endohydrolases', *Mol Gen Genet*, 224 (3), 437-49.

Sokhansandzh, A., et al. (1997), '[Transfer of bacterial genes for proline synthesis in plants and their expression by various plant promotors]', *Genetika*, 33 (7), 906-13.

Spertini, D., Beliveau, C., and Bellemare, G. (1999), 'Screening of transgenic plants by amplification of unknown genomic DNA flanking T-DNA', *Biotechniques*, 27 (2), 308-14.

Springer, N. M., et al. (2007), 'Assessing the efficiency of RNA interference for maize functional genomics', *Plant Physiology*, 143 (4), 1441-51.

Sundaresan, V. and Ramachandran, S. (2001), 'Transposons as tools for functional genomics', *Plant Physiology and Biochemistry*, 39 (3-4), 243-52.

Sussman, M. R., Krysan, P. J., and Young, J. C. (1999), 'T-DNA as an insertional mutagen in Arabidopsis', *Plant Cell*, 11 (12), 2283-90.

Taji, T., et al. (2008), 'Large-scale collection and annotation of full-length enriched cDNAs from a model halophyte, Thellungiella halophila', *BMC Plant Biol*, 8, 115.

Toyoda, T. and Shinozaki, K. (2005), 'Tiling array-driven elucidation of transcriptional structures based on maximum-likelihood and Markov models', *Plant Journal*, 43 (4), 611-21.

Tran, L. S. and Mochida, K. (2010), 'Functional genomics of soybean for improvement of productivity in adverse conditions', *Funct Integr Genomics*, 10 (4), 447-62.

Tsoi, T. V., et al. (1988), '[Cloning and expression of Pseudomonas putida gene controlling the catechol-2,3-oxygenase activity in Escherichia coli cells]', *Genetika*, 24 (9), 1550-61.

Tuskan, G. A., et al. (2006), 'The genome of black cottonwood, Populus trichocarpa (Torr. & Gray)', *Science*, 313 (5793), 1596-604.

Umezawa, T., et al. (2008), 'Sequencing and analysis of approximately 40,000 soybean cDNA clones from a full-length-enriched cDNA library', *DNA Res*, 15 (6), 333-46.

Venter, J. C., et al. (2001), 'The sequence of the human genome', *Science*, 291 (5507), 1304-51.

Verbruggen, N. and Hermans, C. (2008), 'Proline accumulation in plants: a review', *Amino Acids*, 35 (4), 753-9.

Verdel, A., et al. (2009), 'Common themes in siRNA-mediated epigenetic silencing pathways', *International Journal of Developmental Biology*, 53 (2-3), 245-57.

Weigel, D., et al. (2000), 'Activation tagging in Arabidopsis', *Plant Physiol*, 122 (4), 1003-13.

Weisshaar, B., et al. (1998), 'Towards functional characterisation of the members of the R2R3-MYB gene family from Arabidopsis thaliana', *Plant Journal*, 16 (2), 263-76.

Weiste, C., et al. (2007), 'In planta ORFeome analysis by large-scale over-expression of GATEWAY-compatible cDNA clones: screening of ERF transcription factors involved in abiotic stress defense', *Plant Journal*, 52 (2), 382-90.

Wheeler, D. A., et al. (2008), 'The complete genome of an individual by massively parallel DNA sequencing', *Nature*, 452 (7189), 872-6.

Xia, Y., et al. (2004), 'An extracellular aspartic protease functions in Arabidopsis disease resistance signaling', *Embo Journal*, 23 (4), 980-8.

Xiang, Y., Huang, Y., and Xiong, L. (2007), 'Characterization of stress-responsive CIPK genes in rice for stress tolerance improvement', *Plant Physiol*, 144 (3), 1416-28.

Xiong, L. Z., et al. (2006), 'Overexpressing a NAM, ATAF, and CUC (NAC) transcription factor enhances drought resistance and salt tolerance in rice', *Proceedings of the National Academy of Sciences of the United States of America*, 103 (35), 12987-92.

Xu, D. Q., et al. (2008), 'Overexpression of a TFIIIA-type zinc finger protein gene ZFP252 enhances drought and salt tolerance in rice (Oryza sativa L.)', *FEBS Lett,* 582 (7), 1037-43.

Yamaguchi-Shinozaki, K., et al. (2006), 'Functional analysis of rice DREB1/CBF-type transcription factors involved in cold-responsive gene expression in transgenic rice', *Plant and Cell Physiology,* 47 (1), 141-53.

Yamamoto, Y. Y., et al. (2003), 'Gene trapping of the Arabidopsis genome with a firefly luciferase reporter', *Plant Journal,* 35 (2), 273-83.

Yang, S., et al. (2010), 'Narrowing down the targets: towards successful genetic engineering of drought-tolerant crops', *Mol Plant,* 3 (3), 469-90.

Yu, H., et al. (2008), 'Activated expression of an Arabidopsis HD-START protein confers drought tolerance with improved root system and reduced stomatal density', *Plant Cell,* 20 (4), 1134-51.

Yusibov, V. M., et al. (1991), 'Phenotypically normal transgenic T-cyt tobacco plants as a model for the investigation of plant gene expression in response to phytohormonal stress', *Plant Mol Biol,* 17 (4), 825-36.

Zamore, P. D., et al. (2000), 'RNAi: double-stranded RNA directs the ATP-dependent cleavage of mRNA at 21 to 23 nucleotide intervals', *Cell,* 101 (1), 25-33.

Zhang, J. Z. (2003), 'Overexpression analysis of plant transcription factors', *Curr Opin Plant Biol,* 6 (5), 430-40.

Zhang, X. D., et al. (1996), 'Expression of the isopentenyl transferase gene is regulated by auxin in transgenic tobacco tissues', *Transgenic Research,* 5 (1), 57-65.

Zhao, Y., et al. (2001), 'A role for flavin monooxygenase-like enzymes in auxin biosynthesis', *Science,* 291 (5502), 306-9.

Zheng, J., et al. (2010), 'Genome-wide transcriptome analysis of two maize inbred lines under drought stress', *Plant Mol Biol,* 72 (4-5), 407-21.

Zhou, J., et al. (2007), 'Global genome expression analysis of rice in response to drought and high-salinity stresses in shoot, flag leaf, and panicle', *Plant Mol Biol,* 63 (5), 591-608.

Zuo, J., et al. (2002), 'The WUSCHEL gene promotes vegetative-to-embryonic transition in Arabidopsis', *Plant Journal,* 30 (3), 349-59.

3

Arabinogalactan Proteins in
Arabidopsis thaliana Pollen Development

Sílvia Coimbra and Luís Gustavo Pereira
University of Porto, Faculty of Sciences, Biology Department and BioFIG
Portugal

1. Introduction

Pollen ontogeny is an attractive model to study cell division and differentiation. The progression from proliferating microspores to terminally differentiated pollen is characterized by large-scale repression of early program genes and the activation of a unique late gene-expression program in mature pollen.

Among the genes, or gene families that conform to the transition from a sporophytic type of development to a gametophytic program are the arabinogalactan protein (AGP) genes. AGPs are a class of plant proteoglycans, virtually present in all plant cells and in all plant species, from Algae to Angiosperms. They are predominantly located at the periphery of cells, i.e. on the plasma membrane and in the apoplast. Such ubiquitous presence insinuates that AGPs are vital components of the plant cell. Indeed, many studies have implicated AGPs in important biological phenomena, such as cell expansion, cell division, cell death, seed germination, pollen tube growth and guidance, resistance to infection, etc. (Seifert & Roberts, 2007). Evidences implicating AGPs in sexual reproduction have been obtained in our group, for several plant species (Coimbra & Duarte, 2003; Coimbra & Salema 1997; Coimbra et al., 2005) but how these molecules exert their function or how they interact with other cell components is yet to be defined. Only then the fragmentary knowledge that we have today about the function of these proteins may become pieces of a puzzle.

Following studies of pollen and pistil development in *Arabidopsis* with anti-AGP monoclonal antibodies (Coimbra et al., 2007; Pereira et al., 2006), it became clear that some AGPs may be suitable as molecular markers for gametophytic cell differentiation. Despite the tissue-specific carbohydrate epitopes of AGPs, these investigations do not allow the study of single AGP gene products. Therefore, a reverse genetics approach was undertaken to try to identify particular phenotypic traits attributable to certain AGPs, namely AGP6 and AGP11 that we had shown earlier to be pollen-specific (Pereira et al., 2006).

2. Arabinogalactan protein structure

2.1 Protein core

AGPs are complex macromolecules composed of a highly glycosylated protein core whose total mass may amount to only 5% or less of the total mass of the molecule (Fig. 1; Bacic et al., 2000; Nothnagel, 1997; Serpe & Nothnagel, 1999).

The nascent polypeptide chains of AGPs follow the cell's secretory pathway, and therefore contain an N-terminal signal sequence which targets the protein to the endoplasmic reticulum (ER). Most AGPs either contain or are predicted to contain, a C-terminal glycosylphosphatidylinositol (GPI) anchor addition sequence that is excised upon transfer of the protein to a pre-formed anchor present in the ER membrane. The mature protein core is rich in Pro/Hyp, Ala, Ser and Thr, which compose specific repetitive sequence modules. The pattern of these modules presumably constitutes the code that accounts for the characteristic glycan chains present in AGPs (Kieliszewski & Shpak, 2001; Schultz et al., 2004; Showalter et al., 2010). The repetitive amino acid modules (such as AP, TP, SP, or combinations of these) are typically scattered throughout the sequence of the mature protein. It is thus not likely that point mutations should affect the function of AGPs. Even if a mutation disrupts a glycosylation site, a number of others remain.

The properties of the protein backbone define subclasses of AGPs (Schultz et al., 2004). "Classical" AGPs typically contain the central Pro/Hyp-rich domain sandwiched between the N-terminal signal peptide and the C-terminal GPI addition sequence; the lysine-rich AGPs are like "classical" AGPs but contain a Lys-rich module, arabinogalactan (AG) peptides contain small mature protein cores, typically under 30 amino acid residues, the fasciclin-like AGPs (FLAs) may be considered chimeric AGPs because they contain typical glycosylation AGP modules and fasciclin-like domains (Johnson et al., 2003; Schultz et al., 2004; Showalter et al., 2010). Other atypical or chimeric AGPs, such as AGP31, have also been described (Showalter et al., 2010).

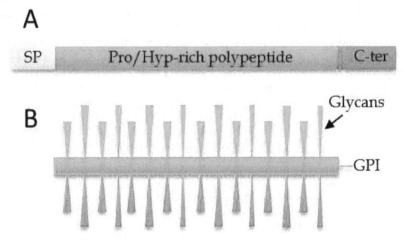

Fig. 1. Highly schematic representation of the molecular structure of a "classical" AGP. (A) Immature polypeptide, and (B) Mature proteoglycan with attached GPI anchor. SP, N-terminal signal peptide; C-ter, C-terminal GPI anchor addition sequence.

2.2 Glycan

AGPs belong to a wider superfamily of plant proteins, the hydroxyproline-rich glycoproteins (HRGPs) which also comprise the extensins and the proline-rich proteins. This classification is mainly based on the sugar content and glycosylation patterns (Kieliszewski

& Shpak, 2001; Shpak et al., 1999) of the molecules, the AGPs being the most heavily glycosylated of all.

AGP glycans are polyssacharide chains O-glycosidically linked to Hyp residues. Polysacharide chains are type II arabinogalactan chains, which consist of a main chain of β-(1→3)-galactopyranose units variously substituted at C6 with oligosacharide or polysacharide chains rich in Gal and Ara but may also contain other monosacharide residues such as Rha and GlcA (see Ellis et al., 2010 for a recent review).

One of the main tools to study the biology of AGPs has been the use of monoclonal antibodies that bind to AGP-specific sugar epitopes. From those studies we know that different AGP epitopes are only present in specific cell types or tissues, or in particular developmental stages. The genetic regulation of this diversity, both in space and in time, is further complicated by the regulation of the necessary battery of glycosyl transferase activities needed to the synthesis of complex sugar chains.

2.3 GPI anchor

The GPI anchor is a post-translational modification of eukaryotic proteins. Newly synthesized proteins that contain a C-terminal GPI-addition signal sequence become attached to a preassembled GPI anchor present in the membrane of the ER, concomitant with the cleavage of the signal sequence. As a result of this modification, GPI-anchored proteins (GAPs) become tethered to the outer layer of the plasma membrane facing the extracellular environment.

Many AGPs were either experimentally shown to be GPI-anchored (about twenty AGPs in *Arabidopsis*; Borner et al., 2003; Elortza et al., 2003, 2006; Lalanne et al., 2004; Schultz et al., 2004) or, based on amino acid sequence analysis, predicted to be GPI-anchored. In *Arabidopsis*, only a few AGPs are not predicted to contain a GPI anchor.

GPI anchors are themselves complex structures, having a highly conserved core composed of ethanolamine-PO_4-6Manα1-2Manα1-6Manα1-4GlcNα1-6*myo*-inositol-1-PO_4-lipid. As opposed to the majority of known GPI anchors, whose lipid part is diacylglycerol-based, the inositolphospholipid part of the only two plant GPI anchors characterized to date is an inositolphosphoceramide (Oxley & Bacic, 1999; Svetek et al., 1999).

Despite the high biological investment in such structures, involving more than twenty dedicated gene products for its biosynthesis, the role of the GPI anchor, or its contribution to the protein biological function is not evident, and remains unsolved for most, if not all, studied examples of GAPs. Moreover, during transit through the secretory system, GPI anchors may be subjected to lipid and/or carbohydrate side-chain remodeling, resulting in a number of GPI structural variants, which may be present in the same organism (Ferguson et al., 2008), and which undoubtedly raises the level of complexity of the biology of GAPs.

Despite all the unanswered questions, specific and commonly referred properties of GAPs mediated by the GPI are their likely association with lipid rafts (Borner et al., 2005), cellular polar sorting (particularly in animal and yeast cells; Legler et al., 2005), and controlled release to the extracellular matrix through the specific action of phospholipases. Indeed, stress conditions such as salinity, cold, drought, heat, wounding, and pathogen attack, are known to activate phospholipase D (PLD), or phospholipase C (PLC) pathways (Testerink & Munnik, 2011), but the actual release of GAPs as a direct consequence of environmental or physiological stimuli is yet to be demonstrated.

Nevertheless, the combination of AGP characteristics, namely cell and/or tissue localization, possible release from the plasma membrane upon stimuli, and complex sugar content that makes AGPs, or specific fragments of AGPs, as likely candidates to perform signaling functions in plants.

3. Pollen development

In flowering plants, development of the haploid male gametophytes (pollen grains) occurs inside a specialized structure called the anther. Successful pollen development, and thus sexual reproduction, requires the correct development of the anther wall layers, and an increased growth of the four locules with fusion into two pollen sacs, and relies on the provision of nutrients and other materials from a specialized secretory tissue, the tapetum. The importance of the tapetum for pollen development is highlighted by findings that the majority of male-sterility mutants involve injuries that affect the expression of tapetum-specific genes. The gene products are released from the tapetal cells and transferred to the pollen surface in wild-type plants. Although there have been many excellent ultrastructural studies of tapetal and pollen development, the underlying biochemical processes have, until recently, remained unclear. This situation is now changing rapidly with the advent of the genome sequence and tools to allow the analysis of gene function (Wilson & Zang, 2009).

3.1 Microsporogenesis

In *Arabidopsis*, at the beginning of anther development, under the protoderm layer, groups of cells develop to give rise to a primary parietal layer and to a sporogenous layer (Owen & Makaroff, 1995). The first will divide to originate the different anther wall layers, and the sporogenous tissue will give rise to the microsporocyte cells. At the pre-meiotic stage of microsporogenesis, the five wall layers of the anther are well differentiated; the microsporocytes have thin cell walls and are successively surrounded by the tapetum, median layer, endothecium and epidermis (Fig. 2).

At the beginning of meiosis, microsporocytes are connected with the tapetal cells by plasmodesmata and it is at this moment that callose deposition begins. During meiosis, the microsporocytes are interconnected by cytomictic channels and their cytoplasm starts to dedifferentiate, which is probably related with the transition from a type of sporophytic gene expression to one of gametophytic gene expression. This transition relates to the change from a diploid to a haploid generation. At this time callose deposition continues, resulting in thick callose walls surrounding the microsporocytes. It has already been assumed that this callose wall can be the trigger to initiate the gametophytic type of development (McCormick, 1993). This physical isolation is important to activate such dramatic changes in development. Following meiosis and cytokinesis, the four haploid microspores are arranged in tetrads encased by a thick callose wall, within the callose wall a microspore-produced cell wall, the primexine is present (Coimbra et al., 2007).

At the end of meiosis, the external walls of the tetrads are dissolved to release individual microspores, by a mixture of enzymes containing endoglucanases and exoglucanases secreted by the tapetum (callase). Alterations in the timing of this event, or failure to express β-1,3-glucanases, leads to abnormal disruption of the callose walls, which has been shown to be a primary cause of male sterility in cytoplasmic male-sterile lines of several species, including Petunia (Izhar & Frankel, 1971).

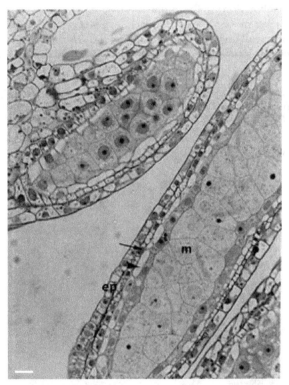

Fig. 2. Light microscopy image of *Amaranthus hypochondriacus* anthers, showing well differentiated microsporocytes (m) and the undifferentiated anther wall layers; from the outside the epidermis (ep), the endothecium (arrow), the middle layer (arrow head) and the tapetum (t) surrounding the microsporocyte cells. Bar 10 μm.

The coordinated development of the microsporocytes and the surrounding cell layers might involve cell-to-cell interactions and localized signaling. From studies with the *SPOROCYTELESS (SPL)* mutants, it was suggested that microsporocyte formation and anther wall development might be coupled. In these mutants, neither microsporocytes nor anther walls are formed and because the microsporocytes and the haploid spores of the tetrads become isolated by callose deposition it is likely that the signaling occurs early during the formation of microsporocytes. The expression of the *SPL* gene in the microsporocytes and its absence from the parietal cells suggests that SPL functions within the microsporocytes to regulate expression of a subset of genes required for microsporocyte formation. Therefore, the authors suggested that the microsporocytes promote, through signal exchange or cell-to-cell interactions, the differentiation and growth of the parietal cell layers, and consequently anther wall development (Yang et al., 1999).

3.2 Microgametogenesis

Microgametogenesis starts with the release of the microspores from the tetrads and leads through a simple cell lineage, to the formation of the mature pollen grains. The ultrastructural description of this developmental stage shows high exocytic activity in tapetum cells, the elaboration of the intine wall by microspores and the deposition of sporopolenin, one of the most complex polymers of plants, which is the outer pollen wall constituent. This outer wall, the exine, is initiated with the formation of the primexine layer at the tetrad stage of development (Owen & Makaroff, 1995).

Microspores start to build a roundish shape due to the formation of a central vacuole, which originates from the fusion and enlargement of small vacuoles. In response to this, the microspore nucleus shifts to an eccentric position against the microspore wall. The first mitosis of the microspore is asymmetric and originates a large and transcriptionally active vegetative cell, and a small generative cell with condensed chromatin that will divide again and originate the two male gametes (Tanaka, 1997) (Fig. 3).

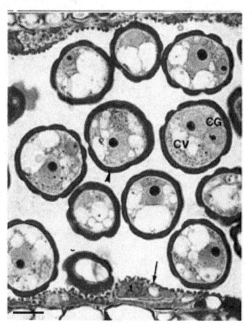

Fig. 3. Anther of *Amaranthus hypochondriacus* showing bicellular pollen grains with a large vegetative cell (CV) and a small and generative cell (CG) in an asymmetric position. The exine wall (arrow head) is being built by the pollen grain and by the sporopolenin depositions from tapetal cells (arrow). Bar 5 μm

The vegetative cell cytoplasm shows strong metabolic alterations related to the asymmetric cell division, one of the most striking events of cell differentiation occurring during the plant life cycle. A single mitotic division will give rise to two completely different cells in size, function and gene activity (Fig. 4). This asymmetric division is vital for generative cell differentiation. Microgametogenesis is quite regular in flowering plants, being the distinction between the timing of the second mitosis the biggest difference; in the majority of Angiosperms, pollen is shed in a bicellular condition, and the generative cell divides inside the pollen tube; whereas in tricellular pollen species, like *Arabidopsis*, the generative cell divides inside the anther. Among the first gametophytic cell markers to be identified is the tomato pollen-specific LAT52 promoter that drives expression specifically in the vegetative cell after pollen mitosis I (Eady et al., 1995), and the lily generative cell-specific H3 histones (gcH2A, gcH3) (Xu et al., 1999). The cytoskeleton plays a central role in determining both nuclear migration and the eccentric division plane. After mitosis, the generative cell migration is a uniquely specialized cell-cell support, which creates a "cell within a cell". The generative cell then assumes a spindle shape and DNA replication takes place, whereas the vegetative nucleus remains arrested in G1 (Fig. 5E). At the end of microgametogenesis, the exine wall will be completely formed, with depositions secreted by tapetal cell (Fig. 3) and the tapetal cells will finally degenerate, depositing their contents in the exine sculptures of the pollen grains, forming the pollen coat.

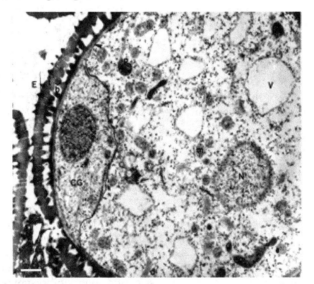

Fig. 4. Electron micrograph of a young bicellular pollen grain of *Amaranthus hypochondriacus*. The generative cell (CG) just formed is still in a spindle shape adjacent to the pollen wall. E-exine (b- bacullae, v– vacuole, m- mitochondria, n- nucleus).

All the steps of this complex process rely on a complex network of signaling events, which probably involve molecules of different kinds (Preuss, 2002). Pollen ontogeny is an attractive model to study cell division and differentiation. In this genomic era, new technologies begin to unravel the roles of specific genes involved in male gametogenesis (Honys et al., 2006; Quan et al., 2008; Toller et al., 2008; Twell 2011).

3.3 Pollen tube growth

Pollen tube growth begins with its emergence from the vegetative cell of the pollen grain after adhesion to the stigma surface. This process depends on the hydration of the pollen grain which is accomplished after recognition. A strictly apical cell growth process maintains the pollen tube cytoplasm and its cargo, the sperm cells, in the most proximal region of the tube as it elongates through the female sporophytic tissues. Callose plugs are laid down at regular intervals behind the growing tip, and the region adjacent to the plug becomes vacuolated. The extreme end of the growing tip consists of a highly dynamic clear zone that contains vesicles and cell wall precursors. Activity in this zone involves continual biosynthesis of cell wall and plasma membrane and turnover of cytoskeletal components as the tube elongates.

In recent years, substantial progress has been made toward understanding tip growth mechanisms (Feijó et al., 1995). Pollen tube reorientation can occur in several minutes, so this might exclude the involvement of newly synthesized gene products. The existence of a tip-focused gradient of cytosolic free calcium has been shown to be maintained by an asymmetric activity of calcium channels and to be essential for growth (Malhó et al., 1995). Polymerization of the actin cytoskeleton is also essential, and tip localized F-actin is thought to mediate the membrane trafficking of secretory vesicles to the apex.

Mutants have been described in which pollen tubes are unable to locate the ovules, as a result of altered interactions with these structures (Higashiyama et al., 2003; Hülskamp et al., 1995). There has been controversy over whether a diffusible signal attracts the pollen tube or whether the female tissues define its path, but recent genetic and physiological data for several plant species, has showed that the female gametophyte produces at least one diffusible signal, which is derived from the two synergids cells (Higashiyama et al., 2003).

4. Arabinogalactan proteins in *Arabidopsis* pollen development

4.1 Immunolocalization of glycan epitopes

Specific monoclonal antibodies (mAbs) that bind to structurally complex carbohydrate epitopes of AGPs have been very useful in revealing the developmental dynamics of the AGP glycan moiety and represent a diagnostic tool for AGPs. Frequently used anti-AGP mAbs are JIM8, JIM13, JIM14, LM2, and MAC207. Accumulated information, obtained by the extensive use of anti-AGP mAbs by the scientific community, shows that AGPs are finely regulated and differentially expressed during pollen development, namely during sexual plant reproduction.

Given the importance of *Arabidopsis* as a model plant, a detailed map of AGP sugar epitopes in different flower parts and at different stages of development was obtained and clearly showed differences in the pattern of distribution of specific AGP sugar epitopes, during *Arabidopsis* anther development. These differences are apparent both in sporophytic and in gametophytic tissues, and it became evident that AGP-specific epitopes can work as markers for certain cell or tissue types, in very precise stages of sporogenesis and gametogenesis.

In the premeiotic stage of microsporogenesis, the epitopes recognized by the mAbs JIM8 and JIM13 are specifically and intensely localized in the tapetum cells and in microsporocytes (Fig. 5A). The selective labelling of the microsporocyte walls is quite important at this stage of development when microsporocytes start a dedifferentiation program related with the transition from a sporophytic gene expression to a type of gametophytic gene expression. This transition relates to the change of generations, from diploid to haploid. As soon as prophase starts, the callose deposition also starts, resulting in thick callose walls surrounding the microsporocytes (Fig. 5C). After meiosis and cytokinesis of the four haploid microspores, the tetrads are completely encased by a thick callose wall. Within the callose wall a microspore-produced cell wall, the primexine, is present. It is interesting to notice that the first existent separation wall in microsporocytes is still present at this stage and still labelled by mAbs JIM8 and JIM13. This labelling pattern can be associated to the signals that must be produce for the efficient release of callase, by the tapetum cells, or a type of developmental time specificity related to the gametophytic development (Fig. 5D).

The presence of AGPs in the tapetum clearly shows that this tissue synthesizes and secretes these molecules (Fig. 5D). The interaction of the sporophytic tapetal cells and the gametophytic differentiation of meiocytes into microspores, are present in the synchronism of callase release from the tapetum ER, as well as from the sporopolenin precursors released into the anther locule. AGPs are strong candidates for cell differentiation signals at this stage of development. It can also be correlated the presence of AGPs with tissues that are set out to programmed cell death. In this study it was observed that the stage at which programmed cell death is triggered, the end of the tetrad stage, is associated with the stronger presence of AGPs recognized by mAbs JIM8 and JIM13.

The labelling with mAbs JIM8 and JIM13 is also strong in the microspores outer surface, which is the site where the intine wall will be built, indicating some association of this important developmental stage with AGP synthesis (Fig. 5D).

During pollen development, it is well documented the strong metabolic alterations in the vegetative cell cytoplasm related to the asymmetric cell division, one of the most striking events of cell differentiation occurring during the plant life cycle. JIM8 and JIM13 specifically label the generative and gametic cells, but not the gametophytic cell. This labelling may function as a molecular marker for cell development and may also be related to the signals necessary to direct these cells inside the pollen tube into their targets, in the embryo sac (Fig. 5E). Moreover, after the second pollen mitosis, the two resulting sperm cells that are inside the pollen grain or inside the pollen tube are still strongly labelled by these two mAbs. The specific labelling of the generative cell was also reported for oilseed rape (Pennell et al., 1991), for Nicotiana tabacum (Li et al., 1995) and for Brassica campestris male gametes (Southworth & Kwiatkowski, 1996). JIM8 and JIM13 do not label the pollen tube wall, which instead is labelled by MAC207 and LM2 (Pereira et al., 2006). These two mAbs are probably related to epitopes in structural AGPs present in several types of plant cell walls.

Antibodies MAC207 and LM2 showed similar binding patterns, both defining extended cell populations in different tissues (Figs. 5B and 5F), as opposed to JIM8 and JIM13 which seemed to define single tissues or single cell types and that can be suitable as molecular markers for pollen development in Arabidopsis.

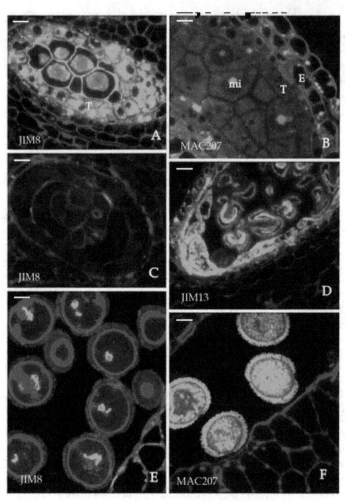

Fig. 5. Fluorescence microscopy of *Arabidopsis* anthers at different stages of pollen development labelled with monoclonal antibodies specific for arabinogalactan proteins (JIM8, JIM13 and MAC207) with FITC-conjugated secondary antibody. A) JIM8 labelling at the stage before meiosis. Tapetal cells and microsporocytes are labelled but there is a remarkable labelling of the wall that surrounds the microsporocytes that are dedifferentiating. B) Same stage of development labelled by MAC207 showing a diffuse labelling in the endothecium and epidermis cells. C) Tetrads of haploid microspores encased by a thick callose wall, the mAb JIM8 is labelling the initial wall of the microsporocytes. D) Microspores just released from tetrads. AGP epitopes recognized by JIM13 are present in tapetal cells and also in the endothecium, which is now developing. This mAb also binds to the microspores cytoplasm and outer surface where the intine wall will eventually develop. E) Tricellular pollen showing specifically the labelling of the two male gametes with JIM8. F) Same stage as in E showing a unspecific labelling with MAC207. T – Tapetum; mi – microsporocytes, E – endothecium.

In order to study the adhesion of pollen tubes via cell wall molecules, the cell wall characteristics of in vitro-grown *Arabidopsis* pollen tubes were investigated using a combination of immunocytochemical and biochemical techniques. Results showed a well-defined localization of cell wall epitopes. Low esterified homogalacturonan epitopes were found mostly in the pollen tube wall back from the tip. Xyloglucan and arabinan from rhamnogalacturonan I epitopes were detected along the entire tube within the two wall layers and the outer wall layer, respectively. In contrast, highly esterified homogalacturonan and arabinogalactan protein epitopes were found associated predominantly with the tip region. This work demonstrated that the *Arabidopsis* pollen tube wall has its own characteristics compared with other cell types in the *Arabidopsis* sporophyte which are probably related to the pollen tube specific growth dynamics (Dardelle et al., 2010).

4.2 AGP tagging

The use of tags engineered into specific AGP gene products is a particularly pertinent approach to study individual AGPs since alternative approaches, such as the use of specific anti-peptide antibodies or even basic analytical tools such as SDS-PAGE may convey challenging difficulties imposed by the massive presence of glycan chains of indeterminate composition and size. Several studies have been published during recent years that include experiments with AGP tagging (Levitin et al., 2008; Li et al., 2010; Sun et al., 2004a, 2004b, 2005; Yang & Showalter, 2007; Van Hengel & Roberts 2003; Zhao et al., 2002). The DNA constructs that underlie those experiments have been used for intracellular localization, expression characterization or purification.

For intracellular localization studies, DNA constructions should contain: i) a functional signal peptide sequence to direct the fusion protein to the ER; ii) the reporter gene, in general green fluorescent protein (GFP) gene has been used; and iii) the mature AGP sequence together with the GPI addition sequence for proper anchoring to the plasma membrane. Constructions such as these have been placed under the influence of the 35S Cauliflower Mosaic Virus (CaMV) promoter for constitutive overexpression studies (phenotype analysis) and to study the consequences of deleting the GPI addition sequence or the Lys-rich domain in Lys-rich AGPs (Li et al., 2010; Sun et al., 2004a, 2004b, 2005; Yang & Showalter, 2007; Zhao et al., 2002). DNA constructions placed under the control of AGP endogenous promoters instead of the 35S CaMV promoter are presently under analysis in our laboratory (unpublished results).

Van Hengel & Roberts (2003) used a *c-myc* tag placed at the C-terminus of a DNA construction containing the complete coding sequence of AGP30 (35S::AGP30-*c-myc*) and used it for electroblot analysis of plant extracts. Indeed, for those few *Arabidopsis* AGPs that may not have a GPI anchor, the tag may be placed at the C-terminus with reduced risk of becoming separated from the remaining polypeptide.

For cell or tissue expression studies, DNA constructions are technically simpler, and may consist of endogenous AGP promoter sequence plus reporter gene, with or without AGP-specific sequences. This approach has been used to confirm and characterize the developmental expression of two pollen-specific *Arabidopsis* classical AGPs (AGP6 and AGP11), and one pollen-specific FLA (FLA3), using either GFP, GUS, or the red fluorescent protein RFP as reporter genes (Coimbra et al., 2008, 2009; Levitin et al., 2008; Li et al., 2010). The temporal expression of AGP6 was determined with great accuracy in *Arabidopsis* plants that were transformed with a ProAGP6::GFP gene construct. GFP fluorescence was absent in all vegetative plant parts, but became clearly visible just after the appearance of the locules in anthers. This stage was identified as corresponding to stage 9 (as described by Smyth et al., 1990). GFP fluorescence was limited to pollen and pollen tubes and could be clearly differentiated from the green yellow autofluorescence characteristic of the exine and of the endothecium lignin thickenings of the anther wall. GFP fluorescence persisted through to the mature pollen grains and was observed in growing pollen tubes (Coimbra et al., 2008; 2009).

4.3 Reverse genetics

The distribution and expression patterns of pollen-specific AGPs have been examined closely in our laboratory. So we tried to identify particular phenotypic traits attributable to either AGP6 or AGP11, or both, in a reverse genetics approach. *AGP6* and *AGP11* are closely related genes, sharing 68% of the amino acid sequence, and therefore seemingly constituting a pair of paralog genes, the function of which may be mutually overlapping. Ds transposon insertion mutant lines for these two genes were available from RIKEN BioResource Center. However, as with so many other plant single gene null mutants, these did not produce recognizable phenotypes (Coimbra et al., 2009). At least in the case of AGP6 and AGP11, a double *agp6 agp11* null mutant did produce identifiable phenotypic traits which could be ascribed to the simultaneous lack of both gene products. In *agp6 agp11* double homozygous mutant lines, but also in plants homozygous for one of the insertions and heterozygous for the other, many of the pollen grains failed to develop normally and collapsed, indicating that the genes are important gametophytically for pollen development (Fig. 6). The collapsed pollen phenotype of *agp6 agp11* was characterized both by scanning electron microscopy and by transmission electron microscopy of pollen grains that clearly showed the degeneration of pollen contents (Coimbra et al., 2009).

Despite the collapsed pollen phenotype typical of the homozygous *agp6 agp11* mutants, a percentage of the pollen grains were able to develop and germinate into functional pollen tubes, as assessed by the presence of seeds in self-pollinated plants. Thus it can be assumed that AGP6 and AGP11 are non-essential for stabilizing pollen grain development, or else that an alternative pathway, potentially involving the ectopic or up-regulation of the expression of other AGP family members, is able partially to compensate for the loss of the two proteins.

In other studies aiming at a more inclusive phenotypic characterization of the *agp6 agp11* double null mutants, it was detected that a number of pollen grains germinated precociously inside the anthers and this phenomenon was dependent upon the relative humidity of the growth chamber (Coimbra et al., 2010). Pollen germination inside anthers was not observed in single *agp6* or *agp11* mutants, and therefore those observations indicated that the double mutation was needed to induce the precocious germination character. Precocious germination of pollen was never found in wild-type plants, even in conditions of high relative humidity, a factor that increased the presence of the phenotypic trait in the double mutant. AGP6 and AGP11 thus seem to have a role in preventing an early and wasteful germination of pollen inside the anthers. As a rule, untimely germination inside anthers does not occur, so some factor or factors must be preventing it from happening. It is also noteworthy that the minimal conditions for pollen germination and tube growth are reproduced inside the anthers. The fact that pollen tubes can germinate and elongate inside the anther locules poses interesting questions regarding germination control and nutritional requirements necessary to support the high respiration rates generally believed to occur for rapid tube growth. It is indeed interesting and challenging to appreciate that arabinogalactan proteoglycans may be interfering with the timing of pollen germination, maybe by a relatively simple process of modulating access of water for hydration (or for the earlier dehydration process), or by interfering with some kind of signaling pathway.

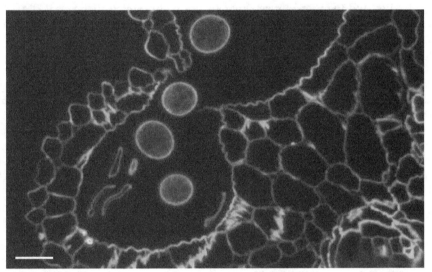

Fig. 6. Light micrograph of an *agp6 agp11* anther showing collapsed pollen grains and some roundish normal pollen grains.

4.4 Microarrays

A series of studies have revealed distinctive transcriptome profiles in microspores, mature pollen and germline cells. The most complete datasets have been generated in the genetic model *Arabidopsis*, where transcriptome analyses have revealed developmental expression profiles from microspores to mature pollen (Honys & Twell, 2004; Pina et al., 2005), the sperm cell transcriptome (Borges et al., 2008) and transcript changes associated with pollen germination and tube growth (Quin et al., 2009; Wang et al., 2008). Sequentially, in our work with the aim of dissecting the biological function of AGPs, we decided to evaluate the whole set of pollen tube expressed genes. For this purpose, we performed microarrays, using the Affymetrix ATH1 genome arrays in the *agp6 agp11* double null mutant pollen tube. We believe that this work is of great general interest for the field of plant science, not only because it highlights interactions between AGPs and other specific gene products but also because of the surprisingly high number of genes whose expression is altered in the pollen tubes of the mutant line, revealing that these molecules can only be involved in very complex phenomena.

This work identified 1300 genes which have either reduced or elevated expression in the *agp6 agp11* pollen tube as compared to wild-type pollen tube, being some of these genes completely inactivated. These genes can be used as starting points to dissect the gene regulatory networks where AGPs are involved during pollen tube growth.

5. Arabinogalactan proteins in other species pollen development

5.1 Rice

Working with *Oryza sativa indica*, Anand & Tyagi (2010) reported the molecular characterization and the promoter activity of a rice pollen-preferential gene, OSIAGP. The authors isolated this gene and identified it as an arabinogalactan protein gene, during a differential screening of inflorescence-specific cDNA libraries. OSIAGP protein has a secretory domain at its N-terminus and *in silico* analysis revealed it to be a secretory and transmembrane protein. When databases were searched using OSIAGP nucleotide and protein sequences, they showed significant homology with a pollen-preferential gene, AGP23 from *Arabidopsis* (AT3G57690). The eight amino acids of the secretory domain at the N-terminus are 100% conserved among OSIAGP and AGP23. The AGP23 gene is 369 bp long and encodes a 61-amino acid protein with homology with other arabinogalactan proteins from *Arabidopsis*. OSIAGP is 59 amino acids and a pollen-preferential gene falling in the category of late pollen genes and is speculated to play important role in pollen tube growth. Its promoter harboring regulatory elements for pollen expression and light regulation could be of interest to the plant community.

Ma & Zhao (2010) identified 69 AGPs from the rice genome, including 13 classical AGPs, 15 AG peptides, 3 non-classical AGPs, 3 early nodulin-like AGPs (eNod-like AGPs), 8 non-specific lipid transfer protein-like AGPs (nsLTP-like AGPs), and 27 FLAs. The results from expressed sequence tags, microarrays, and massively parallel signature sequencing tags revealed that several rice AGP-encoding genes are predominantly expressed in anthers and display differential expression patterns in response to abscisic acid, gibberellic acid, and abiotic stresses, which is highly in agreement with the microarray results obtained in our group for *Arabidopsis* (manuscript in preparation). The two classical AGP-encoding genes, OsAGP7 and OsAGP10, are highly expressed in pollen, similar to AtAGP6 and AtAGP11. The phylogenetically closest rice gene is OsAGP6, but it has a different expression pattern from that of AtAGP6 and AtAGP11. In this case, it is possible that the genes sharing the same function are not those with higher identity. Therefore, OsAGP7 and OsAGP10 may play a conserved role in pollen development, like AtAGP6 and AtAGP11 which redundantly control pollen development and fertility (Coimbra et al., 2009, 2010; Levitin et al., 2008).

5.2 Brassica

The firstly characterized pollen-specific putative AGP genes, Sta 39-4 and Sta 39-3 were isolated from *Brassica napus* flowers in 1996 (Gerster et al., 1996). These two genes are highly homologous; they are 95% identical at the nucleic acid level and 98% identical at the amino acid level. Park et al. (2005) isolated and characterized a pollen preferential gene, *BAN102*, from the Chinese cabbage (*B. campestris*). After analyzing its sequence by BLAST search they found that the coding region of *BAN102* gene had great similarity with AGP23 gene from *Arabidopsis*. The similarities of nucleotide and amino acid sequences were 91% (170/186) and 90% (55/61), respectively.

The *BcMF8* (*B. campestris* male fertility 8) gene, possessing the features of classical AGP was later isolated from *B. campestris*. This gene was highly abundant in the fertile flower buds but silenced in the sterile ones of genic male sterile A/B line ('ZUBajh97-01A/B'). Expression patterns analysis suggested *BcMF8* to be a pollen-specific gene, whose transcript started to be expressed at the uninucleate stage and maintained throughout up to the pollination stage. Isolation and multiple alignments of the homologs of BcMF8 gene in the family *Cruciferae* indicated that *BcMF8* was highly conserved in this family sharing high sequence identity with those of the putative pollen-expressed AGPs genes *Sta 39-4* and *Sta 39-3*, and a lower similarity with that of AGP genes *AtAGP11* and *AtAGP6*.

Besides *Sta 39-4*, *Sta 39-3*, *AtAGP11*, *AtAGP6*, and *BAN102*, a pollen-specific AGP gene *PO2* from alfalfa has been characterized by Qiu et al. (1997) using a similar differential screening technology; after performing an alignment of the deduced BcMF8 protein sequence and its homolog from family *Cruciferae* with all these pollen-specific AGPs mentioned above and the other putative AGPs from the *Arabidopsis* database, the authors found that BcMF8 clustered with those known pollen-specific AGPs, *Sta 39-4*, *Sta 39-3*, *AtAGP11* and *AtAGP6*. This result provided further evidence to the hypothesis that *BcMF8* was pollen-specific. Interestingly *BAN102* and *AtAGP23* are closely related to one another. Indeed, AtAGP23 belongs to AG-peptides subclass which differs in sequence composition from classical AGPs. These results may suggest that pollen development requires different members of the AGP family (Huang et al., 2008).

5.3 Nicotiana

Mollet et al. (2002) reported that *Lycopersicon pimpinellifolium, Aquilegia eximia*, and *Nicotiana tabacum* were not labeled with MAb JIM13 at their tube tips nor did the Yariv reagent bind there and arrest pollen tube growth, as opposed to *Lilium longiflorum and Annona cherimola*. The authors stated that the presence or absence of AGPs at the tube tip appeared to be species dependent. However, they do not exclude the possibility that other AGPs which do not possess the epitopes recognized by these antibodies and/or are not bound with β-GlcY may be present at the pollen tube tips of these species.

Qin and coauthors (Qin et al., 2007), also working with *Nicotiana*, showed that abundant AGPs were present in all areas of the pollen tubes after labeling with JIM13, including the tip region. In pollen tubes, immunogold particles were mainly distributed in the cell wall and cytoplasm, especially around the peripheral region of the generative cell wall. β-D-Glucosyl Yariv reagent, which specifically binds to AGPs, caused slow growth of pollen tubes and reduced immunogold labeling of AGPs with JIM13 *in vitro*. These data suggest that AGPs participate in male gametogenesis and pollen tube growth and may be important surface molecules in generative and sperm cells.

5.4 Malus

By searching for anther-specific genes involved in male gametophyte development in apple (*Malus* x *domestica* Borkh. cv. Fuji) by differential display-PCR, three full-length cDNAs were isolated, and the corresponding genomic sequences were determined by genome walking. The identified genes were intron-less with 228- to 264-bp open reading frames and shared 82–90% nucleotide sequence. Sequence analysis identified that they encoded a putative AGP and were designated *MdAGP1*, *MdAGP2*, and *MdAGP3*, respectively. By RT (reverse transcriptase)-PCR the authors showed that the *MdAGP* genes were selectively expressed in stamens. Promoter analysis confirmed that the MdAGP3 promoter was capable of directing anther or pollen specific expression of the GUS reporter in tobacco and apple. Furthermore, expression of ribosome-inactivating protein under the control of the MdAGP3 promoter induced complete sporophytic male sterility as expected (Choi et al., 2010).

6. Mode of action of AGPs; What are the possibilities?

As we have shown, most of the evidence involving AGPs in pollen development is based on immunolocalization of sugar epitopes, in the genetic manipulation of individual AGP backbone peptides or in the binding of AGPs to β-Yariv reagent, a synthetic reagent which binds AGPs in general, perturbing its biological function. Although significant to find modes of action for the AGPs, these observations are currently difficult to interpret. Also important is to compare the biology of AGPs with that of analogous molecules in animals, trying to match up to their modes of action.

6.1 Arabinogalactan proteins as cell wall components

Although precise functions of AGPs remain elusive, they are widely implicated in plant growth and development. When the Yariv reagent is added to seedlings or to cells in culture, a strong inhibition or even a growth arrest is observed, one possibility for this is the involvement of AGPs in cell wall polymer biosynthesis (Seifert & Roberts, 2007).

In plants most cells expand by diffuse growth; whereas root hairs and pollen tubes expand by tip growth. It is now well documented the high expressed number of AGP genes in the root hair transcriptome (Johnson et al., 2003) and on the pollen tube expressed genes profile (Wang et al., 2008). AGPs are deposited into the tip of growing pollen tubes (Dardelle et al., 2010; Pereira et al., 2006) and tip-growing pollen tubes are inhibited by genetic interference with GPI-anchor biosynthesis (Lalanne et al., 2004) and also by β-Yariv reagent treatment, leading to ectopic callose deposition and pectin alteration (Mollet et al., 2002). These results together entail the involvement of AGPs in growth and development.

The presence of AGP carbohydrate epitopes and AGP backbone peptides in secondary wall thickenings suggests that AGPs might be secreted to the cell surface in parallel with cellulose synthase, and they might be released from their GPI anchor and incorporated into cell wall thickening (Seifert & Roberts, 2007).

A well-characterized classical tomato AGP containing a glycosylphosphatidylinositol plasma membrane anchor sequence was used to elucidate functional roles of AGPs.

Transgenic tobacco BrightYellow-2 (BY-2) cells stably expressing GFP-LeAGP-1 were plasmolysed and used to localize LeAGP-1 on the plasma membrane and in Hechtian strands. Cytoskeleton disruptors and β-Yariv reagent were used to examine the role of LeAGP-1 as a candidate linker protein between the plasma membrane and the cytoskeleton. This study used two approaches. First, BY-2 cells, either wild-type or expressing GFP-microtubule (MT)-binding domain, were treated with β-Yariv reagent, and effects on MTs and F-actin were observed. Second, BY-2 cells expressing GFP-LeAGP-1 were treated with amiprophosmethyl and cytochalasin-D to disrupt MTs and F-actin, and effects on LeAGP-1 localization were observed. Collectively, these studies indicated that GPI-anchored AGPs function to link the plasma membrane to the cytoskeleton (Sardar et al., 2006).

6.2 Arabinogalactan proteins as gradients

Plant reproduction involves a series of interactions between the male gametophyte (the pollen grain or pollen tube) and the extracellular matrix molecules secreted by different cell types along the pollen tube growth pathway in the pistil. These interactions are believed to signal and regulate the pollen tube growth process to effect successful delivery of the sperm cells to the ovules where fertilization takes place. AGPs are believed to play a broad range of functions, ranging from providing structural integrity to mediating cell-cell interactions and communication. Upon germination on the stigma, pollen tubes elongate in the stylar transmitting tract, aided by female factors, with speed and directionality not mimicked in *in vitro* pollen tube growth cultures. It was shown that a stylar transmitting tissue arabinogalactan protein from *N. tabacum*, a TTS (Transmitting Tissue Specific) protein, stimulates pollen tube growth *in vivo* and *in vitro* and attracts pollen tubes grown in a semi-*in vivo* culture system. Within the transmitting tissue, TTS proteins display a gradient of increasing glycosylation from the stigmatic end to the ovarian end of the style, coincident with the direction of pollen tube growth (Cheung et al., 1995; Wu et al., 1995).

Gradients of morphogens are a hallmark in animal development, chemoattractants are important to microbial as well as to animal cell motility systems and developmental pathways; they are also believed to function in the directional growth of pollen tubes (Reger et al., 1992). The gradient of increasing TTS protein glycosylation coincident with the direction of pollen tube elongation is a unique protein-based sugar gradient observed in the female reproductive tissue.

6.3 Arabinogalactan proteins as signaling molecules

AGPs have frequently been hypothesized to be sources of soluble signal molecules, in the form of sugar chain fragments, it is now well established that sugars act as signaling molecules (Hanson & Smeekens 2009) and/or as sources of lipid signal molecules, in the form of the lipid chains that are liberated upon fracture of the GPI anchor, and which may diffuse in the plane of the plasma membrane. Despite the scarcity of experimental evidence to support such roles, it is indeed the combination of AGP main features, namely cell surface localization, likely presence in membrane lipid rafts through their GPI anchor, possible controlled release from the plasma membrane upon stimuli, and the complexity of their sugar content, that makes AGPs, or specific fragments of AGPs, as likely candidates to perform signaling functions in plants, and in plant reproductive processes, in particular. In mammals, species specificity in fertilization occurs through the interactions between sperm cell and egg cell, with cell surface proteins acting as key determinants (Vieira & Miller, 2006). Although flowering plants and mammals have evolved very divergent mechanisms for fertilization and reproductive species recognition, it is highly likely that in both cases GAPs may be used to regulate key steps in fertilization phenomena.

PTs with their unique type of growth restricted to the tip and with an intense secretion and endocytic activity needed for this rapid cell expansion at the apex are always depending on the communications with the pistil molecules. Most probably, the AGPs released to the membrane as part of this exocytic activity are fundamental pieces in the signaling network that directs with precision, pollen tubes to their target, the embryo sac cells.

7. Conclusions and future perspectives

Experiments performed integrating reverse genetics and other experimental approaches, led us to believe that the observed fertility reduction in *agp6 agp11* double null mutant was due to abortion of pollen grains during development (Coimbra et al., 2009). We have further characterized the anthers and pollen of *agp6 agp11* and concluded that both AGPs, AGP6 and AGP11, are necessary for the proper pollen tube growth as well as for preventing untimely pollen grain germination (Coimbra et al., 2010). Further details on the biology of pollen-specific AGPs are expected to emerge from a microarray experiment performed on RNA isolated from pollen tubes of *agp6 agp11* mutant, and which is currently under analysis.

Whether AGPs are predominantly structural, or nutrient-providers, or signaling molecules, is yet to be determined. We are committed to search for AGP-specific ligands recently identified for tobacco pistil AGPs (Lee et al., 2008), reason why we performed microarrays in the double null mutant, *agp6 agp1* pollen tubes, to try to bring some clarification for the biological way of action of this ubiquitous class of plant proteoglycans.

Key future challenges are also the elucidation of the enzymatic machinery that synthesizes the AG carbohydrate structure and the molecular nature and biological role of endogenous AG-specific carbohydrate hydrolases. AGP galactosyltransferase (GalT) activities in tobacco and *Arabidopsis* microsomal membranes were studied with an *in vitro* GalT reaction system. This *in vitro* assay reported to detect GalT activities using AGP peptide and glycopeptide acceptor substrates provides a useful tool for the identification and verification of AGP-specific GalT proteins/genes and an entry point for elucidation of arabinogalactan biosynthesis for AGPs (Liang et al., 2010).

8. Acknowledgment

This work was supported by funding from FCT (Fundação para a Ciência e Tecnologia, Portugal) within the project PTDC/AGR-GPL/67971/2006.

9. References

Anand, S & Tyagi, A. K. (2010). Characterization of a pollen-preferential gene OSIAGP from rice (Oryza sativa L. subspecies indica) coding for an arabinogalactan protein homologue, and analysis of its promoter activity during pollen development and pollen tube growth. *Transgenic Research*, 19, 385–397.

Bacic, A., Currie, G., Gilson, P., Mau, S.-L., Oxley, D., Schultz, C.J., Sommer-Knudsen, J. & Clarke, A.E. (2000). Structural classes of arabinogalactan-proteins. In: *Cell and Developmental Biology of Arabinogalactan-Proteins*, E.A. Nothnagel, A. Bacic and A.E. Clarke (Eds.), Kluwer Academic Publishers/Plenum, Dordrecht, Netherlands/New York, pp. 11–23.

Borges, F., Gomes, G., Gardner, R., Moreno, N., McCormick, S., Feijó, J. & Becker, J. (2008). Comparative Transcriptomics of Arabidopsis Sperm Cells. *Plant Physiology*, 148, 1168–1181.

Borner, G. H. H., Lilley, K. S., Stevens, T. J. & Dupree, P. (2003). Identification of glycosylphosphatidylinositol-anchored proteins in Arabidopsis. A proteomic and genomic analysis. *Plant Physiology*, 132, 568–577.

Borner, G. H. H., Sherrier, D. J., Weimar, T., Michaelson, L. V., Hawkins, N. D., MacAskill, A., Napier, J. A., Beale, M. H., Lilley, K. S. & Dupree, P. (2005). Analysis of detergent-resistant membranes in Arabidopsis. Evidence for plasma membrane lipid rafts. *Plant Physiology*, 137, 104–116.

Cheung, A. Y., Wang, H., & Wu, H.-m. (1995). A floral transmitting tissue-specific glycoprotein attracts pollen tubes and stimulates their growth. *Cell*, 82, 383-393.

Choi, Y-O., Kim, S-S., Lee, S., Kim, S., Yoon, G-B., Kim, H., Lee, Y-P., Yu, G-H., Hyung, N-I. & Soon-Kee Sung, S-K. (2010). Isolation and promoter analysis of anther-specific genes encoding putative arabinogalactan proteins in Malus x domestica. *Plant Cell Reports*. 29, 15–24.

Coimbra, S. & Duarte, C. (2003). Arabinogalactan proteins may facilitate the movement of pollen tubes from the stigma to the ovules in *Actinidia deliciosa* and *Amaranthus hypochondriacus*. *Euphytica*, 133, 171-178.

Coimbra, S., Almeida, J., Junqueira, V., Costa, M. & Pereira, L. G. (2007). Arabinogalactan proteins as molecular markers in Arabidopsis thaliana sexual reproduction. *Journal of Experimental Botany*, 58, 4027–4035.

Coimbra, S., Almeida, J., Monteiro, L., Pereira, L. G. & Sottomayor, M. (2005). Arabinogalactan proteins as molecular markers for generative cell differention and development in Arabidopsis thaliana. *Acta Biologica Cracoviense*, 47 supp. 1, 36.

Coimbra, S., Costa, M. L., Jones, B. J., Mendes, M. A. & Pereira, L. G. (2009). Pollen grain development is compromised in Arabidopsis agp6 agp11 null mutants. *Journal of Experimental Botany*, 60, 3133–3142.

Coimbra, S., Costa, M. L., Mendes, M. A., Pereira, A., Pinto, J. & Pereira, L. G. (2010). Early germination of Arabidopsis pollen in a double null mutant for the arabinogalactan protein genes AGP6 and AGP11. *Sexual Plant Reproduction, 23*, 199–205.

Coimbra, S., Jones, B. J. & Pereira, L. G. (2008). Arabinogalactan proteins (AGPs) related to pollen tube guidance into the embryo sac in Arabidopsis. *Plant Signaling & Behavior, 3*, 455–456.

Coimbra, S., Salema, R. (1997). Immunolocalization of arabinogalactan proteins in *Amaranthus hypochondriacus* L. ovules . *Protoplasma,* 199, 75-82.

Dardelle, F., Lehner, A., Ramdani, Y., Bardor, M., Lerouge, P., Driouich, A. & Mollet, J-C. (2010). Biochemical and Immunocytological Characterizations ofArabidopsis Pollen Tube Cell Wall. *Plant Physiology,* 153, 1563-1576.

Eady, C., Lindsey, K. & Twell, D. (1995). The significance of microspore division and division symmetry for vegetative cell-specific transcription and generative cell differentiation. *Plant Cell,* 7, 65-74.

Ellis, M., Egelund, J., Schultz, C. J. & Bacic, A. (2010). Arabinogalactan-proteins: Key regulators at the cell surface? *Plant Physiology, 153,* 403-419.

Elortza, F., Mohammed, S., Bunkenborg, J., Foster, L. J., Nuhse, T. S., Brodbeck, U., Peck, S. C. & Jensen, O. N. (2006). Modification-specific proteomics of plasma membrane proteins: Identification and characterization of glycosylphosphatidylinositol-anchored proteins released upon phospholipase d treatment. *Journal of Proteome Research, 5,* 935-943.

Elortza, F., Nuhse, T. S., Foster, L. J., Stensballe, A., Peck, S. C. & Jensen, O. N. (2003). Proteomic analysis of glycosylphosphatidylinositol-anchored membrane proteins. *Molecular & Cellular Proteomics, 2,* 1261-1270.

Feijó, J., Malhó, R. & Obermeyer, G. (1995). Ion dynamics and its possible role during in vitro pollen germination and tube growth. *Protoplasma,* 187, 155-167.

Ferguson, M. A., Kinoshita, T. & Hart, G. W. (2009). Glycosylphosphatidylinositol anchors. In: *Essentials of Glycobiology,* A. Varki, R. D. Cummings, J. D. Esko, H. H. Freeze, P. Stanley, C. R. Bertozzi, G. W. Hart & M. E. Etzler (Eds.). Cold Spring Harbor Laboratory Press. Retrieved from http://www.ncbi.nlm.nih.gov/books/NBK1966/.

Gerster, J., Allard, S., Robert, L.S. (1996) Molecular characterization of two *Brassica napus* pollen-expressed genes encoding putative arabinogalactan proteins. *Plant Physiology,* 110, 1231-1237.

Hanson, J. & Smeekens, S. (2009). Sugar perception and signaling —an update. *Current Opinion in Plant Biology,* 12, 562-567.

Higashiyama, T., Kuroiwa, H. and Kuroiwa, T. (2003). Pollen-tube guidance: beacons from the female gametophyte. *Current Opinion in Plant Biology.* 6, 36 – 41.

Honys, D. & Twell, D. (2004). Transcriptome analysis of haploid male gametophyte development in Arabidopsis. *Gene Biology,* 5, R85.

Honys, D., Oh, S-A., Renak, D., Donders, M., Solcova, B., Johnson, J. A., Boudova, R. & Twell, D. (2006). Identification of microspore-active promoters that allow targeted manipulation of gene expression at early stages of microgametogenesis in Arabidopsis. *BMC Plant Biology,* 6, 31.

Huang, L., Cao, J.-S., Zhang, A.-H & Ye, Y.-Q. (2008). Characterization of a putative pollen-specific arabinogalactan protein gene, BcMF8, from *Brassica campestris* ssp. *Chinensis. Molecular Biology Reports,* 35, 631-639.

Hulskamp, M., Kopczak, S., Horejsr, T. F., Kihl, B. K. & Pruitt, R.E. (1995).Identification of genes required for pollen-stigma recognition in Arabidopsis thaliana. *Plant Journal,* 815, 703-714.

Izhar, S. & Frankel, R. (1971) Mechanisms of male sterility in Petunia: The relationship between pH, callase activity in the anthers and the breakdown of microsporogenesis. *Theoretical and Applied Genetics*, 41, 104-8.

Johnson, K. L., Jones, B. J., Bacic, A. & Schultz, C. J. (2003). The Fasciclin-like arabinogalactan proteins of Arabidopsis. A multigene family of putative cell adhesion molecules. *Plant Physiology*, 133, 1911-1925.

Kieliszewski, M. J. & Shpak, E. (2001). Synthetic genes for the elucidation of glycosylation codes for arabinogalactan-proteins and other hydroxyproline-rich glycoproteins. *Cellular and Molecular Life Sciences*, 58, 1386-1398.

Lalanne, E., Honys, D., Johnson, A., Borner, G. H. H., Lilley, K. S., Dupree, P., Grossniklaus, U. & Twell, D. (2004). SETH1 and SETH2, two components of the glycosylphosphatidylinositol anchor biosynthetic pathway, are required for pollen germination and tube growth in Arabidopsis. *The Plant Cell*, 16, 229-240.

Lee, C. B., Swatek, K. N. & McClure, B. (2008). Pollen proteins bind to the C-terminal domain of Nicotiana alata pistil arabinogalactan proteins. *Journal of Biological Chemistry*, 283, 26965-26973.

Legler, D. F., Doucey, M.-A., Schneider, P., Chapatte, L., Bender, F. C. & Bron, C. (2005). Differential insertion of GPI-anchored GFPs into lipid rafts of live cells. *The FASEB Journal*, 19, 73-75.

Levitin, B., Richter, D., Markovich, I. & Zik, M. (2008). Arabinogalactan proteins 6 and 11 are required for stamen and pollen function in Arabidopsis. *The Plant Journal*, 56, 351-363.

Li, J., Yu, M., Geng, L.-L. & Zhao, J. (2010). The fasciclin-like arabinogalactan protein gene, FLA3, is involved in microspore development of Arabidopsis. *The Plant Journal*, 64, 482-497.

Li, Y.Q., Faleri, C., Geitmann, A., Zhang, H-Q. & Cresti M. (1995). Immunogold localization of arabinogalactan proteins, unesterified and esterified pectins in pollen grains and pollen tubes of *Nicotiana tabacum*. *Protoplasma* 189, 26-36.

Liang, Y., Faik, A., Kieliszewski, M., Tan, L., Wen-Liang Xu, W-L. & Showalter, A. M. (2010). Identification and Characterization of in Vitro Galactosyltransferase Activities Involved in Arabinogalactan-Protein Glycosylation in Tobacco and Arabidopsis. *Plant Physiology*, 154, 632-642

Ma, H & Zhao, J. (2010). Genome-wide identification, classification, and expression analysis of the arabinogalactan protein gene family in rice (Oryza sativa L.). *Journal of Experimental Botany*, 61, 2647-2668.

Malhó, R., Read, N., Trewavas, A. & Pais, M. S. (1995). Calcium Channel Activity during Pollen Tube Growth and Reorientation. *The Plant Cell*, 7, 1173-1184

McCormick, S. (1993). Male gamethophyte development. *The Plant Cell*, 5, 1265-1275.

Mollet, J. C., Kim, S., Jauh, G. Y. & Lord, E. M. (2002). Arabinogalactan proteins, pollen tube growth, and the reversible effects of Yariv phenylglycoside. *Protoplasma*, 219, 89-98.

Nothnagel, E.A. (1997). Proteoglycans and related components in plant cells. *International Review of Cytology*, 174, 195-291.

Owen, H. & Makaroff, C. (1995). Ultrastructure of microsporogenesis and microgametogenesis in *Arabidopsis thaliana* (L.) Heynh. Ecotype Wassilewskija (Brassicaceae). *Protoplasma*, 185, 7-21.

Oxley, D. & Bacic, A. (1999). Structure of the glycosylphosphatidylinositol anchor of an arabinogalactan protein from pyrus communis suspension-cultured cells. *Proceedings of the National Academy of Sciences*, 96, 14246-14251.

Park, B. S., Kim, J. S., Kim, S. H. & Park, Y. D. (2005). Characterization of a pollen-preferential gene, BAN102, from Chinese cabbage. *Plant Cell*, 14, 1-8.

Pennell, R. I., Janniche, L., Kjellbom, P., Scofield, G. N., Peart, J.M. & Roberts, K. (1991). Developmental regulation of a plasma membrane arabinogalactan protein epitope in oilseed rape flowers. *Plant Cell*, 3, 1317-1326.

Pereira, L. G., Coimbra, S., Oliveira, H., Monteiro, L. & Sottomayor, M. (2006). Expression of Arabinogalactan Protein Genes in Pollen Tubes of *Arabidopsis thaliana*. *Planta*, 223, 374-380.

Pina, C., Pinto, F., Feijó, J. & Becker, J. (2005). Gene family analysis of the Arabidopsis pollen transcriptome reveals biological implications for cell growth, division control, and gene expression regulation. *Plant Physiology*, 138, 744-756.

Preuss, D. (2002). Sexual signalling on a cellular level: lessons from plant reproduction. *Molecular Biology of the Cell*, 13, 1803-1805.

Qin, Y., Leydon, A. R., Manziello, A., Pandey, R., Mount, D., Denic, S., Vasic B., Johnson, M. & Palanivelu, R. (2009). Penetration of the Stigma and Style Elicits a Novel Transcriptome in Pollen Tubes, Pointing to Genes Critical for Growth in a Pistil. *PLoS Genetics*, 5, e1000621.

Qin, Y., Chen, D. & Zhao, J. (2007). Localization of arabinogalactan proteins in anther, pollen, and pollen tube of Nicotiana tabacum L. *Protoplasma*, 231, 43-53.

Qiu, X., Wu, Y. Z., Du, S. & Erickson, L. (1997). A new arabinogalactan protein-like gene expressed in the pollen of alfalfa. *Plant Science*, 124, 41-47.

Quan, L., Xiao, R., Li, W., Oh, SA., Kong, H., Ambrose, J. C,, Malcos, J. L., Cyr, R., Twell, D. & Ma, H. (2008). Functional divergence of the duplicated *AtKIN14a* and *AtKIN14b* genes: critical roles in Arabidopsis meiosis and gametophyte development. *Plant Journal*, 53, 1013-1026.

Reger, B. J., Chaubal, R. & Pressey, R. (1992). Chemotropic responses by pearl millet pollen tubes. *Sexual Plant Reproduction*, 5, 47-56.

Sardar, H. S., Yang, J., & Showalter, A. (2006). Molecular Interactions of Arabinogalactan Proteins with Cortical Microtubules and F-Actin in Bright Yellow-2 Tobacco Cultured Cells. *Plant Physiology*, 142, 1469-1479.

Schultz, C. J., Ferguson, K. L., Lahnstein, J. & Bacic, A. (2004). Post-translational modifications of arabinogalactan-peptides of Arabidopsis thaliana. *Journal of Biological Chemistry*, 279, 45503-45511.

Seifert, G. J. & Roberts, K. (2007). The biology of arabinogalactan proteins. *Annual Review of Plant Biology*, 58, 137-161.

Serpe, M.D. & Nothnagel E.A. (1999). Arabinogalactan-proteins in the multiple domains of the plant cell surface. *Advances in Botanical Research*, 30, 207-289.

Showalter, A. M., Keppler, B., Lichtenberg, J., Gu, D. & Welch, L. R. (2010). A bioinformatics approach to the identification, classification, and analysis of Hydroxyproline-Rich glycoproteins. *Plant Physiology*, 153, 485-513.

Shpak, E., Leykam, J. F., & Kieliszewski, M. J. (1999). Synthetic genes for glycoprotein design and the elucidation of hydroxyproline-O-glycosylation codes. *Proceedings of the National Academy of Sciences*, 96, 14736-14741.

Smyth, D. R., Bowman, J. L. & Meyerowitz, E. M. (1990). Early flower development in Arabidopsis. *The Plant Cell, 2,* 755-767.

Southworth, D. & Kwiatkowski, S. (1996). Arabinogalactan proteins at the surface of *Brassica* sperm and *Lilium* sperm and generative cells. *Sexual Plant Reproduction, 9,* 296-272.

Sun, W., Kieliszewski, M. J. & Showalter, A. M. (2004a). Overexpression of tomato LeAGP-1 arabinogalactan-protein promotes lateral branching and hampers reproductive development. *The Plant Journal, 40,* 870-881.

Sun, W., Xu, J., Yang, J., Kieliszewski, M. J. & Showalter, A. M. (2005). The lysine-rich arabinogalactan-protein subfamily in Arabidopsis: Gene expression, glycoprotein purification and biochemical characterization. *Plant and Cell Physiology, 46,* 975-984.

Sun, W., Zhao, Z. D., Hare, M. C., Kieliszewski, M. J. & Showalter, A. M. (2004b). Tomato LeAGP-1 is a plasma membrane-bound, glycosylphosphatidylinositol-anchored arabinogalactan-protein. *Physiologia Plantarum, 120,* 319-327.

Svetek, J., Yadav, M. P. & Nothnagel, E. A. (1999). Presence of a glycosylphosphatidylinositol lipid anchor on rose arabinogalactan proteins. *Journal of Biological Chemistry, 274,* 14724-14733.

Tanaka, I. (1997). Differentiation of generative and vegetative cells in angiosperm pollen. *Sexual Plant Reproduction, 10,* 1-7.

Testerink, C. & Munnik, T. (2011). Molecular, cellular, and physiological responses to phosphatidic acid formation in plants. *Journal of Experimental Botany, 62,* 2349-2361.

Toller, A., Brownfield, L., Neu, C., Twell, D. & Schulze-Lefert, P. (2008). Dual function of Arabidopsis glucan synthase-like genes *GSL8* and *GSL10* in male gametophyte development and plant growth. *Plant Journal, 54,* 911-923.

Twell, D. (2011). Male gametogenesis and germline specification in flowering plants. *Sexual Plant Reproduction, 24,* 149-160

Van Hengel, A. J. & Roberts, K. (2003). AtAGP30, an arabinogalactan-protein in the cell walls of the primary root, plays a role in root regeneration and seed germination. *The Plant Journal, 36,* 256-270.

Vieira, A. & Miller, D. J. (2006). Gamete interaction: is it species specific? *Molecular Reproductive Development, 73,* 1422-1429.

Wang, Y., Zhang, W. Z., Song, L. F., Zou, J.J., Su, Z. & Wu, W. H. (2008). Transcriptome analyses show changes in gene expression to accompany pollen germination and tube growth in Arabidopsis. *Plant Physiology, 148,* 1201-1211.

Wilson, Z. A. & Zhang, D-B. (2009). From Arabidopsis to rice: pathways in pollen development. *Journal of Experimental Botany, 60,* 1479-1492.

Wu, H.-M., Wang, H., & Cheung, A. Y. (1995). A Pollen Tube Growth Stimulatory Glycoprotein Is Deglycosylated by Pollen Tubes and Displays a Glycosylation Gradient in the Flower. *Cell, 82,* 383-393.

Xu, H., Swoboda, I., Bhalla, P. L. & Singh, M. B. (1999) Male gametic cell specific expression of H2A and H3 histone genes. *Plant Molecular Biology, 39,* 607-614.

Yang, J. & Showalter, A. M. (2007). Expression and localization of AtAGP18, a lysine-rich arabinogalactan-protein in Arabidopsis. *Planta, 226,* 169-179.

Yang, W.C, Ye, D., Xu, J., Sundaresan, V. (1999). The *SPOROCYTELESS* gene of *Arabidopsis* is required for initiation of sporogenesis and encodes a novel nuclear protein. *Genes Development*, 15, 2108–2117.

Zhao, Z. D., Tan, L., Showalter, A. M., Lamport, D. T. & Kieliszewski, M. J. (2002). Tomato LeAGP-1 arabinogalactan-protein purified from transgenic tobacco corroborates the hyp contiguity hypothesis. *The Plant Journal, 31*, 431–444.

4

Transgenic Plants as Biofactories for the Production of Biopharmaceuticals: A Case Study of Human Placental Lactogen

Iratxe Urreta and Sonia Castañón
Neiker-Tecnalia, Vitoria Gasteiz
Spain

1. Introduction

Throughout human evolution, plants have provided us with food, fibers to produce clothes, and medicines to treat different kind of diseases. In fact, we could say that plants represent the "chemists" of the antiquity. Examples of important therapeutic molecules obtained from plants are morphine, atropine, ephedrine, codeine, and digitalin and so on (Farnsworth et al., 1985). Plants have always been a common source of medicaments, either in the form of traditional preparations or as pure active principles. For many people worldwide, natural plant-based remedies are used to treat both acute and chronic health problems. Indeed, about 40% of the drugs prescribed in the USA and Europe come from active compounds found in plants (Rates, 2001; Sivakumar, 2006). The chemical synthesis of these plant-derived compounds led to the production of pharmaceuticals at an industrial level, allowing the development of medicament industry. These active compounds from plants have been basically small molecules. Aspirin, one of the most famous medicaments, was developed as an analogous of the salicylic acid extracted from willow bark in the 19st century (Knäblein, 2005). Moreover, the development of more sophisticated extraction and purification procedures allowed the extraction of alkaloids (morphine) and other kind of molecules from plants and mammals (Liénard et al., 2007). However, chemical synthesis has some limitations to produce complex therapeutic molecules such as antibodies. So, many therapeutic molecules have to be isolated and purified from living material (e.g. from blood plasma), which involves a high risk of transmission of pathogens to the manufactured product (Engelhard, 2007).

The emergence of genetic engineering in the early 1970s has made possible a new way to produce pharmaceuticals outside their natural host. The term "biopharmaceutical" appears to be originated in the 1980s and is used to refer to therapeutic proteins produced by modern biotechnological techniques (Walsh, 2005a). Genetic transformation techniques have allowed the scientist to transform living organisms like bacteria, yeast, animal cells and plants into production "biofactories". These organisms are forced to produce customized therapeutics making possible the treatment of diseases like genetic disorders, AIDS, and diabetes. Unlike chemically produced therapeutics, biopharmaceuticals are our own molecules and hence more compatible with biological systems (Rai & Padh, 2001).

Biopharmaceuticals represent the fastest growing sector within pharmaceutical industry, with worldwide sales of $94 billion from the total market of $600 billion in 2007. In 2009, biopharmaceuticals recorded global sales of $99 billion ($61 billion for therapeutic proteins and $38 billion for monoclonal antibody (mAb)-based products (Walsh, 2010), and these numbers are expected to reach sales of $125 billion by the year 2015 (Xu et al., 2011). Regarding the most lucrative molecules, mAb-based products indicated for treating cancer are the bestsellers of biopharmaceuticals. The next most lucrative group is represented by insulin and insulin analogs, generating $13.3 billion in sales, followed by EPO-based products whose sales stands at $9.5 billion (Walsh, 2010).

The biopharmaceutical industry relies basically in non-human mammalian cell lines like CHO cells, and bacterial systems. In mammalian cell lines human proteins are correctly processed and modified, but the scalability is limited, maintenance of bioreactors is expensive, and there is a risk of contamination of the product with human pathogens (Daniell et al., 2001; Twyman et al., 2005). *Escherichia coli* was the pioneer expression system and in 1977 it was used for the successful expression of somatostatin (growth hormone-inhibiting hormone). *E. coli* is widely used because it is the most cost-effective production system, allowing the large-scale production of proteins (Rai & Padh, 2001). Insulin produced in *E. coli*, also known as "humulin", received the marketing authorization in 1982 which represented the beginning of the biopharmaceutical industry (Walsh, 2005b). However, the prokaryotic nature of this organism limits the complexity of the proteins that can be correctly processed, and the appearance of inclusion bodies increases the cost of the product (Boehm, 2007). Yeast-based systems are able to grow in well-defined simple media, in large-scale bioreactors and at high cell-densities (Gerngross, 2004). Therefore interferon was expressed in yeast (1981). However, the low product yields, inefficient protein secretion and hyperglycosylation of proteins (addition of large number of mannose residues) are common in problems in this system. Transgenic animals allow the scaling-up of protein production but they require long generation times, and have technical difficulties. Moreover, ethical aspects also limit this system (Hunter et al., 2005). Insect cell-based systems provide a eukaryotic expression system readily amenable to scale-up, but they are inefficient in proteolytic cleavage of proteins as well as for glycosylation (Rai & Padh, 2001). Nevertheless, β interferon and factor IX were expressed in insect cells (1983) and transgenic animals (1988), respectively (Desai et al., 2010).

Genetic engineering techniques have made it possible to produce therapeutic molecules using different host system. But what are the ideal characteristics that should satisfy an expression system to be considered a good system? It is important to (1) produce proteins with the correct conformation and biologically active, (2) allow good productivities, (3) have minor economic costs associated to manufacturing and purification, and (4) be recognized as safe system by the society.

Despite the variety of expression systems available for biopharmaceutical industry, none of them meets completely all the requirements for a production system to be able to produce any kind of therapeutic protein. There is no ideal or universal expression system. All of them have some advantages and some limitations, so the choice for the best system should be done by evaluating biological as well as economic aspects, and bearing in mind the purpose of the product of interest.

2. Transgenic plants as production platform for biopharmaceuticals

The total global market for biopharmaceuticals is projected to grow at between 7% and 15% annually over the next several years, with mAb-based approvals continuing to dominate (Walsh, 2010). Efforts on human genome project have identified new proteins with therapeutic potential and the growing number of candidates in clinical trials will be in the market in the next years. So it just keeps getting more apparent that the traditional systems used by the industry so far will not be able to meet the growing demand and will represent a bottleneck in bringing therapeutic proteins to the society. Pharmaceutical industry requires a production platform able to provide enough quantity of safe and high-quality products at the lowest cost. Transgenic plants are becoming one of the most interesting alternative systems for biopharmaceutical production as they offer many advantages over traditional systems.

One of the most important benefits of transgenic plants is the capacity to obtain high production volumes at relatively low cost, due to their flexibility in terms of scale-up (Sparrow et al., 2007). It has been estimated that the cost associated with the production of 300 kg of a secretory antibody in maize or tobacco plants is 0.5-2 million $ per year, while the cost increases to 6-7 million per year using transgenic goat or mammalian cell cultures (Gerlach et al., 2010). Transgenic plants allow the increase or decrease of cultivated area depending on the market demand. Moreover, the availability of harvesting, storage and transport infrastructures turns plants into a very efficient production system (Kusnadi et al., 1997).

As eukaryotic organisms, protein synthesis pathways are highly conserved among plants and animals, so plant cells are able to correctly fold and assemble proteins as well as to perform post-translational modifications required for protein activity and stability (Kamenarova et al., 2005). This is demonstrated by the ability of plant cells to produce various types of antibodies, such as IgGs and IgAs, which are complex proteins requiring the assembly of various polypeptide chains through disulphide bonds (Twyman et al., 2003). Apart from cultured mammalian cells, only plants are able to assembly the light and heavy chains of antibodies (Gomord et al., 2004). For this reason, amino acid sequences of human proteins expressed in plant cells are usually the same as native counterparts, which ensures the quality of products produced in plants (Schillberg & Twyman, 2007). Moreover, plant cells have a very high ratio of biologically active protein (92% of total protein in tobacco BY-2 cells) when compared to *E. coli* (12%) and *P. pastoris* (40%) (Boehm, 2007). Proteins produced in plant cells are less likely to be contaminated with human or animal pathogens and hence product safety can be guaranteed.

The diversity of plant host makes possible the production of oral vaccines in edible parts of the plant, eliminating the need of cold-chain and economic costs associated with purification processes, and thus being more amenable for developing countries. Moreover, tissues like seeds, tubers or fruits allow the storage and stability of biopharmaceuticals produced in plants (Desai et al., 2010).

One of the major challenges of plant systems, besides low protein expression level which will be discussed below is the inability of plants, as in other eukaryotic systems, to perfectly reproduce human-type glycosylation on biopharmaceuticals. Glycosylation is the most widespread post-translational modification with more than half of the human biopharmaceuticals being glycoproteins. Glycosylation affects their function, plasma half-life and/or biological activity (Saint-Jore-Dupas et al., 2007). Main differences between plants and mammal cells are relative to modification of glycans in Golgi apparatus. Plants do not have sialic acid residues present in human glycoproteins, and add α (1,3)-fucose and β (1,2)-xylose residues which are responsible of immunogenic response in human therapy (Chen et al., 2005). However, there are some strategies available to "humanize" recombinant glycoprotein produced in plants. The first one is based on endoplasmatic reticulum (ER) retention of proteins in their biosynthetic pathway bypassing Golgi apparatus. This can be done through the addition on KDEL sequence in the recombinant protein (Gomord & Faye, 2004). The second one relies on inactivation or down-regulation of plant xylosyltransferase and fucosyltransferase using RNA interference technology. Alternatively it is possible to "humanize" glycoproteins co-expressing human glycosyltransferases (Gomord et al., 2004) or by *in vitro* galactosylation of recombinant proteins using mammalian enzymes (Bardor et al., 2003).

Another future challenge for plants is the concern about biosafety and regulatory issues. General public is very reluctant to transgenic crops, especially since the appearance of those resistant to pests or pesticides, and unfortunately transgenic plants for biopharmaceutical production must suffer the same rejection. The main biosafety problems are related to risks associated to human/animal health and to the environment, so it is very important to evaluate and manage carefully those risks through risk assessments (Peterson & Arntzen, 2004). There are many strategies available to avoid those problems: apart from delimitation of physic barriers and strict agricultural management (crop destruction, field cleaning, and crop rotation) to avoid mixing of modified and food/feed crops, the most obvious alternative is the use of non-food/feed crops (tobacco, *Arabidopsis thaliana*, *Phiscomitrella patens*, *Lemna* sp.), or the use of greenhouses/glasshouses where crops are confined and controlled. Biological containment provides a natural and additional barrier to gene flow. Production of biopharmaceuticals in self-pollinating species (rice, wheat, and pea) or engineering other species to make them cleistogamic (self-pollination before flower opening) is interesting for gene containment (Commandeur et al., 2003).

Biotechnological tools have allowed the development of different strategies for gene and protein containment. Chloroplast transformation technology is a very attractive approach as chloroplast genome is maternally inherited in most crops, and gene spread via pollen occurs at low frequency (Svab & Maliga, 2007). The use of male-sterile plants (Gils et al., 2008) or "terminator" technology (also known as GURTs, "Genetic use restriction technologies), which is based on a repressible system to produce plants with non-viable seeds unless the plants are exposed to specific activators (Lee & Natesan, 2006), have also been used to contain transgenes. Other strategies include organ- or tissue-specific expression of transgenes, ER-retention of proteins and inducible promoters (Obembe et al., 2010).

It is clear that taking into account the previously reported concerns regarding health and environmental risks, transgenic plants producing biopharmaceuticals must be regulated strictly, and in fact they are. However, there is a need for the harmonization of international regulation of plants producing biopharmaceuticals in European Union and USA (Gerlach et al., 2010) and to unify the criterions. Plants as production system for biopharmaceuticals comply not only with the strict regulatory requirements covering other GM crops, but also the regulations set out by agencies that oversee the production of pharmaceuticals (Sparrow et al., 2007).

Plants represent a very versatile and plastic expression system to produce biopharmaceutical proteins, because they offer a great variety of strategies. Depending on production needs, product quality, or compliance with certain legal requirements, we can choose between different plant-based systems without the need to use bacteria or animal cells.

The most used strategy for protein production is that based on whole plants. Most biopharmaceuticals are produced in nuclear transformed plants, which are obtained mainly by using soil pathogen *Agrobacterium tumefaciens,* and allow the stable expression of transgenes with the required post-translational modifications. Genetic transformation of chloroplast to obtain transplastomic plants constitutes a promising alternative strategy for protein production due to the important advantages over nuclear transformation. Proteins can achieve high expression levels due to high copy-number of transgene per cell. It has been reported an expression level of up to 31% of total soluble protein (TSP) for an animal vaccine produced in transplastomic tobacco plants (Molina et al., 2004). Chloroplast transformation is based on homologous recombination events so there is no gene silencing (Maliga, 2003). It is possible to integrate multiple genes in operons and, as previously mentioned, this system offers transgene containment through maternal inheritance (Bock & Khan, 2004). However, in the chloroplast it is not possible to carry out post-translational modifications present in the ER limiting the range of proteins that can be produced in this system. Against stable nuclear or chloroplast transformation, plants can be transiently transformed by agroinfiltration with *A. tumefaciens,* biolistic methods and viral vectors. This strategy is very useful for transformation construct verification and to test functionality of recombinant proteins (Fischer et al., 2004).

2.1 Plant cell cultures

As an alternative to whole plants, biopharmaceuticals can be produced using plant cell cultures. Plant cell culture is a very interesting and promising alternative system and has been used for almost two decades for protein production, as well as for secondary metabolite production.

Suspended cell cultures are derived from callus, which are unorganized and generally undifferentiated cell aggregates derived from plant tissues cultured in solid media supplemented with growth regulators. Calli are suspended in liquid media to form a homogeneous suspension. Transgenic suspension cultures can be obtained from wild type callus or suspensions transformed by *A. tumefaciens* or biolistic methods, depending on plant species, but also from transgenic plant tissue such as leaf or stem. With the second approach there is no need for further genetic manipulation like transformed tissue selection and line screening (Hellwig et al., 2004).

Since 1990, when Sijmons et al. reported the expression of human serum albumin in transgenic tobacco suspension cells, a diverse array of biopharmaceuticals have been produced using suspended cell cultures. Huang & McDonald (2009) reported an extensive list of recombinant proteins produced in this system, including human erythropoietin, human granulocyte-macrophage colony-stimulating factor (hGM-CSF), human interleukins, hepatitis B surface antigen, and many types of antibodies. Although productivity of plant cell cultures can vary considerably from 0.5 μg/L to 200 mg/L (Hellwig et al., 2004), many proteins have been produced at high yields (>10 mg/L) (Xu et al., 2011). Most of biopharmaceutical proteins have been expressed in tobacco cells like well characterized BY-2 and NT-1 cell lines, because these host lines are fast-growing, well synchronized and susceptible to *Agrobacterium* mediated transformation (Nagata & Kumagai, 1999). Other host plants used include rice, an emerging host specie (Kim et al., 2008), sweet potato (Min et al., 2006), tomato (Kwon et al., 2003), and carrot (Shaaltiel et al., 2007).

Suspended cell cultures offer many advantages over whole plant systems, and perhaps one of the most important is that *in vitro* plant cultures, as grown in confined and sterile conditions, avoid the political resistance to release of genetically modified plants to the field. Moreover, confined culture eliminates problems related to weather conditions, soil quality, season, plagues, and contamination with agrochemicals and fertilizers that affect field grown plants (Weathers et al., 2010). This is an ideal characteristic for production of high purity biopharmaceuticals. Whole plant systems require longer period of time to be productive (sowing, growing, harvesting) while proteins could be manufactured in days or weeks on a time-scale compatible with that of market demands (Doran, 2000). Growth in bioreactors not only allows the reduction in production cycles, but also the scalability and the tight control of growth parameters (temperature, O_2 supply, agitation, pH), ensuring product quality and batch-to-batch consistency, as well as product traceability (Ma et al., 2005). These features make the system amenable to good manufacturing processes (GMP), facilitating compliance with regulatory and environmental requirements (Spök et al., 2008). Bioreactor technology provides a great variety of culture modes and bioreactor types to increase recombinant protein production. Large-scale culturing can be done in standard stirred-tank bioreactor, pneumatic bioreactor (bubble column or air-lift), wave bioreactor, membrane bioreactor, hollow fiber bioreactor or miniature bioreactor (Huang & McDonald, 2009). Regarding to operation modes, we can choose between batch culture, fed-batch culture, continuous and semi-continuous culture, and perfusion culture (Xu et al., 2011). Choosing a suitable bioreactor type and operation mode should include adequate oxygen mass transfer to cells, low shear stress to cells, and proper nutrient supply to cells and product removal from cells (Huang & McDonald, 2009).

The potential of suspended cells to secrete biologically active proteins into the culture medium has a big impact on downstream processing costs. Cells are grown in relative simple and synthetic protein free media which facilitates the recovery of proteins (Hellwig, 2004). Protein secretion is determined by the presence of signal peptide and protein size to pass through wall pores. Due to their undifferentiated nature, callus cells lack functional plasmodesmata, and as grown in suspended form, there is minimal cell-to-cell communication (Su, 2006). All those features can reduce post-transcriptional gene silencing (PTGS) because signal transmission between cells is avoided (Doran, 2000). In conclusion, plant cell suspension cultures integrate many of the advantages of whole plant systems with those of microorganisms and mammalian cell cultures.

In contrast to suspended cells, plant cells can also be immobilized for biopharmaceutical production. This system is based on the immobilization of cells using encapsulating gels, such as alginate, which protect cells against mechanical damages. Immobilization also facilitates the re-use of cells in continuous or semi-continuous culture, allowing higher inoculums than standard methodologies (James, 2001).

Differentiated organs such as hairy roots and shooty teratomas have also been developed for protein production. Hairy root cultures have been used for decades to produce secondary metabolites such as resveratrol (Medina-Bolivar et al., 2007). To generate hairy roots, wounded transgenic host plant is co-cultivated with *Agrobacterium rhizogenes* which transfers *rol* and *aux* genes which are responsible of root phenotype and induction (Sivakumar, 2006). Hairy root cultures, grown in hormone-free medium, are genetically stable over time and allow uniform expression of proteins at high level in relatively short periods of time (Franconi et al., 2010). Secretion of proteins is also allowed in root cultures which can be cultured in bioreactors or in hydroponic tanks from where they take water and nutrients while releasing proteins continuously (Knäblein, 2005). Shooty teratomas are generated from transgenic seedlings co-cultured with *A. tumefaciens* strain T37. After co-culture and Agrobacteria elimination, shoots are cultured in liquid medium (Sharp & Doran, 2001).

2.2 Emerging plant-based systems

In the last decade alternative novel expression systems have been developed for the production of biopharmaceuticals. We would like to highlight the potential of novel systems such as mosses, algae and aquatic plants which can be cultured in contained conditions in bioreactors.

Mosses as multicellular eukaryotic organisms can produce biologically active therapeutic proteins. *Physcomitrella patens* is the model organism which has been studied for a long time. It can be cultured during its complete lifecycle, but when cultured in liquid medium vegetative growth is favored (Franconi et al., 2010). The fact that *P. patens* is grown in small plant fragments, and not as protoplast, provides genetic stability avoiding somaclonal variation (Knäblein, 2005). The moss is photoautotrophic and only requires inorganic salts, water and CO_2 for growth, so it is easily cultured in stirred glass tanks or tubular photobioreactors. The most interesting feature of *P. patens* is its high frequency of homologous recombination which facilitates the precise knockout of genes (Decker & Reski, 2007). In this context, genes for $\alpha(1,3)$-fucosyltransferase and $\beta(1,2)$-xylosyltransferase were disrupted by homologous recombination to obtain double knockout clones of *P. patens* with no allergenic N-glycans (Koprivova et al., 2004). The potential of this species for biopharmaceutical production gave rise to the company Greenovation in 1999. Since 2001, the company is developing the "bryotechnology", using double-knockout strain of *P. patens* cultured in photobioreactors. Currently this company is producing a variety of biopharmaceuticals such as growth factor (VEGF), serum proteins (HSA), peptide hormones (EPO), enzymes (phosphatase), vaccines and a wide range of oncology mAb´s in 100L tubular bioreactors and 200L disposable bag systems (www.greenovation.com).

Microalgae represent a diverse group of prokaryotic (cyanobacteria) and eukaryotic photosynthetic microorganisms that are found in marine and freshwater environments. Microalgae combine simple and inexpensive growth requirements and ability for post-transcriptional processing of proteins, with the rapid growth rate and potential for high-density culture (Walker et al., 2005). Most green algae are classified as generally regarded as safe (GRAS), making processing of expressed products more amenable to regulatory issues (Potvin & Zhang, 2010), and offering a safe platform for vaccine production. Microalgae can be grown in contained bioreactors in a matter of weeks from initial transformation event to large-scale protein production, and as single cell type culture, there should be less variation in protein accumulation, making downstream processing more uniform (Specht et al., 2010). However, light utilization and distribution in the bioreactor represents a limiting factor for cell growth which requires careful design of bioreactors. In this regard, heterotrophic growth using glucose as carbon source should be advantageous for well established "dark" bioreactors allowing high cell density (Franconi et al., 2010). *Chlamydomonas reinhardtii* is the most successfully used microalgae, because it is genetically well characterized with all three genomes (the nuclear, chloroplast and mitochondrial) sequenced, and genetic trans-formation methods are well established (Rasala & Mayfield, 2011). Proteins such as human antibodies (Mayfield & Franlikn, 2005), human glutamic acid decarboxylase 65 (hGAD65) (Wang et al., 2008), domain 14 of fibronectin (14FN3), VEGF and HMGB1 (Rasala & Mayfield, 2011) have been produced in *C. reinhardtii*.

Finally we would like to pay attention to aquatic higher plants from *Lemnaceae* family such as those from *Lemna*, *Spirodela* and *Wolffia* genus. These edible plants are safe, fast-growing, and easy to grow and harvest species amenable to genetic transformation using *A. tumefaciens* or biolistic method (Weathers et al., 2010). *Lemna minor* has been used to produce mAb´s fused with an RNA interference construct targeting the expression of endogenous genes α(1,3)-fucosyltransferase and β(1,2)-xylosyltransferase to obtain an antibody without plant specific N-glycans (Cox et al., 2006). Many therapeutic proteins have been expressed in this system, mainly by Biolex Company and its subsidiary LemnaGene who produce interferon for hepatitis C treatment, plasmin for thrombosis treatment and anti-CD20 antibody optimized for the treatment of non-Hodgkin's B-cell lymphoma (www.biolex.com). At the present time, *Spirodela oligorrhiza*, with an expression level for GFP protein of 25% of TSP represents the best expressing system for nuclear transformation in higher plants (Franconi et al., 2010).

2.3 Factors influencing therapeutic protein production in plants

Many authors agree that one of the most important limitations of transgenic plants is their low expression level of recombinant proteins. Increasing the amount of recombinant protein is crucial for the system to be economically viable. The yield of recombinant proteins produced in plants depends on many factors: the intrinsic limitations of host plant, limitations imposed by transgene expression and protein stability, which are optimized by careful design of expression vector, and downstream processing. Finally, we would like to analyze the environmental factors affecting transgenic plants in open field.

2.3.1 Choice of host plant

The range of plant species amenable to genetic transformation is very wide so there is no an ideal host for molecular farming. There are many factors needed to be taken into account when choosing the host plant: from production factors such as infrastructure availability, storage and distribution cost, or plant productivity, to factors affecting environment and human health and food safety (Schillberg & Twyman, 2007).

Usually host plants are divided into food crops and non-food crops. Tobacco is the most used specie among non-food crops due to its easy of genetic transformation (both nuclear and chloroplast genomes), high biomass yield and seed production, and ability to produce many therapeutic proteins (antibodies, vaccines, cytokines, serum and blood proteins, hormones). Its widespread use has led to the production of low-alkaloid cultivars suitable for oral delivery of vaccines (Tremblay et al., 2010). It has also been reported that the expression of biologically active GM-CSF in commercial sugarcane which is propagated vegetatively from stem pieces offering a "secure" platform for production of recombinant proteins (Wang et al., 2005). Food-crops include seed-crops, vegetables and fruits. Species such as cereals, rice, maize, soybean, or barley allow the expression of recombinant proteins in the seed which enables the long-term storage and the containment of proteins avoiding the exposure of non-target organisms. However seed-crops must go through a flowering cycle to produce seeds, so it is very important to control pollen transfer (Sparrow et al., 2007). Oilseed crops are useful for protein production because the fusion of recombinant proteins to the endogenous protein oleosin allow the easy recovery of recombinant proteins (Stoger et al., 2005). Vegetable crops include potato, worldwide cultivated specie with a very well-developed agricultural infrastructure. Tuber has been used mainly for oral vaccine production using tissue-specific promoter patatin. This vegetative organ offers stability to the recombinant protein. Carrot, alfalfa, lettuce and spinach have also been used for vaccine production because can be consumed raw, because can be grown again after leaf harvest, or because are very easy to scale-up (Sparrow et al., 2007). Fruits such as tomato and banana, more palatable than potatoes for raw consumption, are suitable for oral vaccines because of high biomass yield (tomatoes) and easy distribution (bananas) (Kamenarova et al., 2005).

2.3.2 Design of the expression vector

The expression vector is the vehicle to integrate the gene of interest into the plant genome. As stated before, the final yield of recombinant protein depends on many factors, and most of them can be addressed through inclusion of suitable regulatory sequences in the vector. In this context we can control factors that affect transcription, translation as well as protein accumulation. Figure 1 shows the schematic representation of a general expression vector showing potential regulatory sequences.

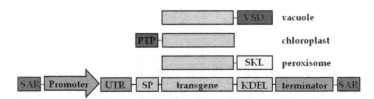

Fig. 1. Schematic diagram of plant expression vector showing the main regulatory sequences. SAR, scaffold attachment region; UTR, untranslated region; SP, signal peptide; KDEL, endoplasmatic retention signal; SKL, peroxisomes target sequence serine-lysine-leucine; PTP, plastid transit peptide; VSD, vacuole sorting determinant.

2.3.2.1 Promoter

This is one of the most important factors affecting transcription of genes because it is responsible of when, where and how does the gene transcribe. Promoters can be from plant or viral origin and usually are divided into constitutive, tissue-specific and inducible categories. Constitutive promoters drive the continuous expression of the transgene in all plant tissues. The best known and most used promoter is derived from the cauliflower mosaic virus (CaMV35S), a strong promoter mainly used for dicot species (Xu et al., 2011).

The duplication of an enhancer region located in the 5' region of CaMV35S promoter allowed the development of the strongest version commonly known as double CaMV35S (2xCaMV35S) (Kay et al., 1987). For monocot species, rice actin gene or bean arcelin gene promoters are most used. The choice of promoter also depends on the nature of recombinant protein: when the protein of interest is toxic for the host plant, tissue-specific expression could be a helpful strategy (Desai et al., 2010). This type of promoter allows the control and restricted expression of transgenes to specific tissues such as leaf (small subunit of Rubisco), seeds (maize globulin-1, barley D-hordein, rice glutelin), and storage organs (tuber patatin). Inducible promoters allow not only the tight control of gene expression, but also the quantity of expression level. Induction can be regulated through chemical stimuli (steroid, sucrose, salt) and/or environmental factors (light, temperature, wounding). In this kind of promoters it is very important to specify between promoter and inductor, the rapid response upon induction, and the safety of the inducer for plant (Corrado & Karali, 2009).

2.3.2.2 5' UTR sequences

The UTR sequences are located in the upstream region of ORF and are related to the efficiency of initiation of translation. In eukaryotes this process is thought to follow the scanning mechanism conducted by small subunit of ribosome from 5'cap of mRNA through the untranslated leader until the first start codon (AUG) is found (Kermode, 2006). Including 5' UTR sequences from Alfalfa mosaic virus (AMV), Tobacco mosaic virus (TMV), *Chalcone synthase* (CHS), or *Alcohol dehydrogenase* (NtADH) have been successfully used to enhance the translation efficiency, allowing transgene levels 30 to 100-fold higher (Satoh et al., 2004; Schiermeyer et al., 2005).

2.3.2.3 Terminator sequences

Also known as polyadenylation signals, these sequences are very important for RNA stabilization because they are responsible for the correct processing of RNA after stop codon (Desai et al., 2010). The most commonly used terminator sequences are derived from the *A. tumefaciens nopaline synthase* gene which has been used successfully in both dicot and monocot plants.

2.3.2.4 Codon optimization of transgenes

Many of the transgenes introduced in plants for biopharmaceutical production come from humans. Codon usage between divergent species is often very different and will affect the expression level of recombinant proteins (Lessard et al., 2002). Based on available data of codon usage for a given host plant, transgene codons with the lowest usage frequency are modified by changing the nucleotide sequence without changing the amino acid sequence (Desai et al., 2010).

2.3.2.5 SAR/MAR sequences

The inclusion of sequences containing scaffold/matrix attachment regions (SAR/MAR) flanking the expression cassette has been used as a mechanism to increase transgene expression. SAR/MAR are AT-rich DNA sequences of 300-500 bp-long which interact with nuclear scaffolds organizing the structure of the genome (Allen et al., 2000). These sequences facilitate transcription of genes by changing the chromatin topology into less condensed regions (Kermode, 2006). With the use of SAR/MAR sequences it is possible to increase transgene expression levels through stabilization of the expression in progeny and reduction of expression variability (Ulker et al., 1999).

2.3.2.6 Subcellular targeting

Specific targeting of recombinant proteins to plant organelles not only guarantees the correct post-translational modifications required, but also allows the enhancement of protein stabilization, minimizing proteolytic degradation and facilitating downstream processing such as purification (Kermode, 2006). Eukaryotic preproteins synthesized with an N-terminal signal peptide are targeted to the secretory pathway. In the absence of signal peptide, recombinant protein is accumulated into the cytosol. In most cases this is not an appropriate compartment for recombinant proteins due to its hydrolytic activity and its negative redox potential which is unfavorable for protein folding (Benchabane et al., 2008).

In the case of proteins with signal peptide and no further targeting, expressed proteins are secreted from the ER to the apoplast. This compartment has been implicated in the production of proteolytic fragments in several transgenic plant systems due to proteases found in the secretory pathway between the ER and Golgi (Doran, 2006).

One of the most used compartments within plant cells for subcellular targeting is the ER. This organelle has many advantages for recombinant protein accumulation due to its specific characteristics. ER has very low hydrolytic activity and can tolerate unusually high accumulation of proteins without compromising plant development and reproduction because of its plasticity to become a reservoir of protein and oil bodies (Vitale & Pedrazzini, 2005). Recombinant proteins targeted to ER are protected from proteolytic degradation and correct folding and disulphide bond formation are allowed (Benchabane et al., 2008). Inclusion of KDEL/HDEL tetrapeptide in the C-terminal region of the protein ensures ER-retention. Targeting recombinant proteins into the ER has led to an increase in expression levels from 4.5 to 100-fold (Fiedler et al., 1997; Schouten et al., 1996; Torres et al., 1999).

Another suitable cell compartment for recombinant proteins accumulation is the vacuole. Important functions of this organelle include control of cell turgor, turnover of macromolecules, sequestration of toxic compounds, and finally, storage of high-energy compounds (Benchabane et al., 2008). From two types of vacuoles in plants, only protein storage vacuoles which are very abundant in seeds are suitable for recombinant proteins because of their mild environment (Stoger et al., 2005).

With chloroplast transformation, it is possible to target the recombinant protein to this organelle through the suitable targeting signals. The chloroplast-transit peptide of potato rbcS1 gene allowed the expression of human papillomavirus type 16 L1 (HPV-16 L1) protein into chloroplast of *Nicotiana benthamiana*, obtaining expression levels of up to 11% of TSP (Maclean et al., 2007). Moreover, chloroplast targeting can avoid the toxic effect of recombinant protein in the cytosol (Gils et al., 2005; Hühns et al., 2008).

It has also been reported that the subcellular targeting to mitochondria (Menassa et al., 2004) and chloroplast and peroxisomes at the same time, it is possible to accumulate the recombinant protein 160% of that in chloroplasts alone and 240% of that in peroxisomes alone (Hyunjong et al., 2006).

2.3.2.7 Protein/tag fusions

Expression of the recombinant protein attached to a protein with enhanced stability may prove to be useful to enhance fusion partner. In plants, several authors have reported the suitability of engineering fusion proteins to improve expression levels (Wirth et al., 2006; Xu et al., 2010). Many fusion proteins can also serve as purification tags due to their specific binding features or easy of recovery, but those will be discussed in the next section.

2.3.3 Downstream processing

The downstream processing of recombinant proteins includes protein extraction, purification and characterization, a process that may represent 80% or more of total production costs (Twyman et al., 2003). Purification strategies may be based on standard chromatographic techniques, but affinity purification using His-tag, protein A, or glutathione S-transferase (GST) is an alternative approach widely used by researchers (Arnau et al., 2006). We have successfully used His-tag to purify human placental lactogen (hPL) from tobacco leaf tissue (Urreta et al., 2010).

Fusion of recombinant proteins to endogenous proteins such as oleosin or γ-zein is also very attractive for therapeutic protein production and is gaining more attention for commercial issues. Oleosin fusion platform has been developed by SemBioSys Inc. (www.sembiosys.com) and allows recombinant protein to be targeted to the oil bodies in rapeseed and safflower. The purification is based on separation of oil bodies by simple and inexpensive schemes (Schilberg & Twyman, 2007). Similarly, fusion of recombinant protein to γ-zein, a prolamin of maize, induces protein body formation and high accumulation of foreign protein within ER. Protein bodies are insoluble and easily purified by centrifugation (Vitale & Pedrazzini, 2005). An alternative emerging purification strategy relies on elastin like polypeptide (ELP) fusion to recombinant protein (Floss et al., 2009).

2.3.5 Environmental factors

Plants, as grown in the field or in the glasshouse, are subjected to environmental factors. Those factors influence their health condition, and subsequently, the quality and the quantity of recombinant protein. The stability of foreign protein in plants grown in the field is very important because it determines the homogeneity of the product. The most important environmental factors are light, temperature, soil (nutrients), water and insect attacks (Jamal et al., 2009). Most of these factors can be controlled to greater or lesser extent for plants cultivated in the glasshouse. However, when transgenic plants are grown in the field, light and temperature became the critical factors. Light is crucial for plants as energy source and temperature also affects plant growth and productivity being especially dangerous at high temperatures. The analysis of the influence of plant´s physiology on recombinant protein accumulation is crucial to plan the best strategy for plant harvesting (Conley et al., 2010).

Although protein production in plants has long been investigated, the most of the reported works are based on plants grown in glasshouses under controlled conditions. Arlen et al. (2007) reported that the field production of chloroplast derived interferon (IFN) from tobacco plants. They cultivated 0.26 acre containing 7369 plants and obtained 107.7 kg of biomass from a single harvest. This biomass contained approximately 87 g IFN (0.8 mg g^{-1}). It must be noted that tobacco plants can be harvested 4-6 times within a grow season from a 1 acre. They also tested light and temperature influence on glasshouse grown plants revealing the importance of leaf maturity and illumination, and most importantly, they observed higher IFN yield in those plants (1-3 mg g^{-1} fresh weigh).

3. Case study: Production of human placental lactogen in transgenic plants

Human placental lactogen (hPL), or chorionic somatomamotropin, is a 22 kDa peptidic hormone secreted by the placenta during pregnancy (Barrera-Saldaña et al. 1982). This protein is involved in the adaptation of islets of Langerhans to pregnancy through the regulation of beta (β) cell mass and function (Brelje et al. 1993). The capacity of hPL to improve β cell function, proliferation and survival *in vitro* and *in vivo* (Vasavada et al. 2000) makes possible its use as therapeutic protein for Langerhans islet transplantation to patients with type 1 diabetes. The transplantation of pancreatic tissue has become an interesting alternative treatment for diabetes, allowing the independence of patients to insulin injection. Unfortunately, the low availability of pancreatic tissue and great prevalence of the disease limits the potential of this treatment. Therefore, hPL together with other peptidic growth factors (HGF and PTHrP) belongs to the short list of proteins capable of improving the critical beta cell parameters, namely function, proliferation and survival, to improve tissue availability for islet transplantation (Fujinaka et al. 2007). Currently, the commercially available hPL protein is purified from human placenta which increases not only the risk of human pathogen propagation but also increases the cost of the protein in the market. This hormone has also been expressed in *E. coli* as inclusion bodies, which required the solubilization and refolding of the protein (Lan et al., 2006), making the process and the product more expensive. Due to the potential of hPL protein for type 1 diabetes treatment, we analyzed the suitability of potato plants as an alternative production system for hPL protein. In our laboratory, we thought that due to the potential of PL as a candidate for the treatment of type-1 diabetes, it would be very interesting to investigate alternative production strategies for the safe and easy production of this protein at low cost. Of course, plants represent for us the better production platform for all the reasons previously exposed and discussed, as well as for our experience on therapeutic protein production in plants.

We have successfully expressed hPL protein in *Nicotiana tabacum* cv. Xanthi plants reaching expression levels of up to 1% of TSP (Urreta et al., 2010). *In vitro* bioassays using the rat insulinoma (INS-1) cell line showed that recombinant protein was able to induce cell proliferation, demonstrating that plant cells can produce the biologically active hPL protein. Due to the difficulty to evaluate the relative performance of different crops for commercial production of therapeutic proteins which requires the production of the same protein in different host plant (Schilberg & Twyman, 2007), we attempted to produce hPL protein in *Solanum tuberosum*. We chose potato because it is easily transformed and widely used for molecular farming. Potato is an important crop worldwide and in our region, Álava (Spain). Indeed, our research institute has a long experience in the cultivation and improvement of this crop, being the reference germplasm bank in Spain.

To achieve our objective, the cDNA encoding h*PL* gene was obtained by RT-PCR from human placenta mRNA (Clontech) as previously described (Urreta et al., 2010). Briefly, the amplified sequence of 654-bp was ligated into pGJ2750 binary vector (kindly provided by Max Planck Institute, Köln, Germany) to obtain pNEKhPL1 expression vector where h*PL* gene is regulated by the promoter and terminator of CaMV 35S gene. The expression vector was introduced into *Agrobacterium tumefaciens* EHA105 strain and co-cultivated with *Solanum tuberosum* cv. Désirée leaf disks following the protocol described by Dietze et al. (1995). The regenerated shoots (Figure 2A) were subcultured in MS medium (Murashige & Skoog 1962) supplemented with kanamycin (100 mg L^{-1}) at 22°C, under 120 µmol m^{-2} s^{-1} of photon flux and a 16-h photoperiod in a growth chamber. Most of the regenerated putative transgenic plants had the same phenotype as the non transformed wild type plants (Figure 2, B-C). Integration of h*PL* and *neomycin phosphotransferase* (*nptII*) genes in the genome of regenerated plants was verified by polymerase chain reaction (PCR) using specific primers. All transgenic plants screened by PCR were positive (Figure 2, D-E), ensuring the correct integration of T-DNA in the genome of plants. Further molecular characterization of recombinant hPL protein was assessed by western blot. The proteins were separated on 12% polyacrylamide gels under reducing conditions and transferred to a nitrocellulose membrane (Bio-Rad, Hercules, CA, USA). The blots were blocked for 1 hour in PBS with 0.05% (v/v) Tween-20 (PBS-T) and 5% non-fat dry milk, and incubated with polyclonal anti-hPL antibody (RB9067, NeoMarkers, Fremont, CA, USA). Blots were incubated with Anti-Rabbit IgG secondary antibody conjugated with alkaline phosphatase (Sigma-Aldrich, St Louis, MO, USA). Antibody binding was detected with NBT/BCIP (Sigma). The recombinant protein was detected in leaf protein extracts of transgenic plants cultivated in the greenhouse, which showed the expected molecular weight of 22 kDa (Figure 2F) demonstrating that plant cells are able to process signal peptides from human origin.

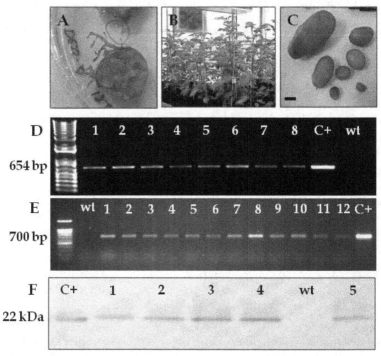

Fig. 2. Phenotypic and molecular characterization of transgenic potato plants. A, regenerated shoots obtained from leaf tissue co-cultured with *A. tumefaciens*. B, phenotype of potato plants cultivated in the glasshouse. C, tubers from transgenic lines (scale 1:2 cm). D and E, PCR amplification of h*PL* (D) and *nptII* (E) genes in the genome of regenerated plants. 1-12 independent lines; c+, pNEKhPL1 vector; wt, non transformed plant. F, western blot of transgenic plants cultivated in the glasshouse. C+, 10 ng of commercial hPL protein (NeoMarkers); 1-5, 20 µg of TSP from different transgenic lines.

The amount of hPL protein was determined by sandwich enzyme linked immunosorbent assay (ELISA) in leaf tissue of plants grown *in vitro* and cultivated in the greenhouse. Maxisorp 96-well microtiter plates were coated with polyclonal anti-hPL antibody (NeoMarkers) overnight at 4°C. The wells were blocked with 1.5% horse serum in PBS-T and then samples as well as serial dilutions of commercially available hPL protein were added in PBS for 1 hour at 37°C. Monoclonal anti-HPL antibody (MCA322, Serotec, Oxford, UK) and alkaline-phosphatase conjugated goat anti-mouse IgG antibody (Sigma) were coated subsequently to finally detect color development using 4-nitrophenyl phosphate (Sigma-Aldrich). Absorbance was measured at 405 nm in a microplate reader (Multiskan RC, Labsystems, Helsinki, Finland) and TSP content was determined by the Bio-Rad protein assay. The results showed that plants cultivated in the greenhouse had 2.4-fold higher expression levels than those grown *in vitro* (Table 1). The lower protein levels detected in plants grown *in vitro* could be due to the stressful conditions of culture conditions. It has been described that the artificial medium and the high air humidity and low gas exchange of this culture type could induce disturbances in plant development (Kadleček et al. 2001) which possibly limit the production of endogenous and foreign protein levels. Using the expression vector pNEKhPL1, the expression levels reached a maximum of 0.21% of TSP. These expression values are similar to those reported by other authors using vectors similar to pNEKhPL1 (Castañón et al., 2002; Kim et al., 2003; Mason et al., 1998; Ritcher et al., 2000).

Culture condition/tissue	N° of plants analyzed	Expression level (mean value±s.e.)	Expression range
In vitro	24	0.25±0.03	0.03-0.63
Glasshouse	24	0.62±0.13	0.06-2.12
Tuber	16	1.34±0.13	0.69-1.87
callus	20	4.46±0.26	2.77-7.41

Table 1. Yield of the recombinant human placental lactogen protein expressed in potato plants grown under different conditions and in different organ/tissues. The expression levels were measured by enzyme-linked immunosorbent assay (ELISA) and are represented as recombinant protein per total soluble protein content (ng hPL μg^{-1} PTS).

Compared to the levels obtained in tobacco plants (Urreta et al. 2010), those reached in potato plants are much lower. However it must be taken into account that pNEKhPL1 expression vector is the simplest version of a plant expression vector; with no ER-retention signal, SAR sequence, nor enhanced 35S promoter included in pNEKhPL2 vector used in tobacco plants (Urreta et al., 2010). This difference in expression levels highlights the importance of the design of expression vector including regulatory sequences suitable to achieve our goals.

Regarding to the influence of host plant, in our laboratory we also transformed tobacco plants with pNEKhPL1 expression vector, obtaining a maximum expression level 2-fold higher than that obtained in potato (data not shown). So in the case of hPL production in plants, tobacco seems to be more suitable than potato with allowing higher expression levels in leaf tissue of plants cultivated in the greenhouse.

Because of the constitutive nature of CaMV35S promoter, we also analysed the expression level of recombinant hPL in tubers from 16 transgenic plants. We obtained elevated levels of recombinant protein in this tissue, with a mean value 2-fold higher than that of leaf tissue

(Table 1). The maximum hPL expression level was 0.18% of TSP, slightly lower than the maximum in leaf tissue. These results are in the range of expression levels reported by other Authors in potato tubers using CaMV35S promoter (Mason et al., 1998; Zhou et al., 2003; Bielmet et al., 2003). Although CaMV35S promoter is not tuber specific, the expression levels obtained are also similar to those reported by Mason et al. (1996) and Castañón et al. (2002) using tuber specific promoters such as patatin. It is also noteworthy that from the tubers tested, 4 of them did not show detectable levels of hPL.

In order to assess the suitability of hPL production in plant cell cultures we induced the formation of callus from leaf tissue of transgenic plants cultivated *in vitro* (Figure 3, A). Leaf explants were cultured in MS medium supplemented with casein hydrolysate (0.2% w/v), 2,4-D (5 mg L^{-1}) and kinetin (0.2 mg L^{-1}), and incubated at 23°C in the dark. Friable callus developed in 3-4 weeks (Figure 3, B) which were analysed by western blot. Recombinant protein was detected at 22 kDa, as expected, but not in all lines tested (Figure 3, C).

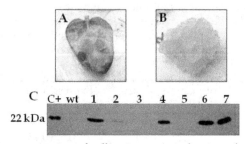

Fig. 3. Induction and characterization of callus tissue. A, induction of callus from leaf tissue cultured in MS medium supplemented with 2,4-D and kinetin. B, isolated callus line from leaf tissue and subcultured to fresh medium. C, western blot of callus protein extracts (10 µg). c+, commercial hPL protein (30 ng) (NeoMarkers); 1-7, transgenic lines; wt, callus induced from leaf explants of non transformed potato plants.

Although those calli were kanamycin-resistant, they did not express hPL protein. This could be as a result of promoter methylation during callus induction process as observed by other Authors (De Carvalho et al., 1992; Fojtova et al., 2003). The hPL protein levels were measured by ELISA. Results showed that a mean value 7-fold higher than that of leaf tissue of plants grown in the glasshouse, reaching a maximum level of 0.74% of TSP (Table 1). The expression of hPL protein in callus is favored because of the lower TSP content in this tissue. Data presented in this work represents a good starting point to further analyze the suitability of cell suspensions as an alternative plant-based expression system for hPL protein production. As previously discussed, plant cell cultures offer many advantages over whole plants for the production of human therapeutics, like rapid growth and easier purification strategies when the protein is secreted to the culture media. The expression vector pNEKhPL1 allows the targeting of hPL to the apoplast and hence to the culture medium. Moreover, using bioreactors we can avoid the release of genetically modified organisms to the field, and improve protein yields through the control of growth parameters allowing a homogeneous batch to batch production.

Finally, to further characterize the transgenic potato plants expressing the hPL protein, we analyzed the stability of the recombinant protein in the second generation (vegetative) of four transgenic lines. Recombinant hPL levels were measured by ELISA, showing that one of the lines lost the expression of hPL and in the remaining lines the levels ranged from 0.2 to 0.38 ng μg^{-1} of TSP. These levels are higher than those of their counterpart grown *in vitro*, and similar to that of first generation cultivated in the glasshouse.

All the results obtained in this work suggest that although hPL protein is correctly produced in potato leaf tissue, tubers accumulate higher mean protein levels, allowing the storage of the protein. Further investigation will be necessary to analyze the stability of the protein in long term storage and the influence of temperature conditions needed to maintain its integrity. Although potato plants show lower expression levels than tobacco plants transformed with the same expression vector, expression levels in callus tissue opens the possibility for the production of hPL protein using plant cell cultures.

4. Conclusion

Transgenic plants represent an attractive production platform for therapeutic proteins due to all of the previously mentioned features. However, the commercial production of plant-made proteins is still limited, as compared to other expression systems currently available. Twenty two years after the first antibody produced in plants, the great research efforts of public and private institutions have lead to hundreds of plant-made protein publications and patents. Plant-based systems have demonstrated their suitability for protein production, ensuring their capacity for correct production of many types of therapeutic proteins and their great versatility to allow the production of proteins in different conditions. Many companies are exploiting plant systems for protein production (SemBiosys, Protalix, Dow Agrosicences, Méristem Therapeutics) but the pharmaceutical industry still shows reluctance to integrate plant systems in their production strategy. The lower protein levels obtained in plant systems can be greatly improved through the careful design of expression vectors and the choice of host plant. On the other hand, there is no biosafety and/or regulatory issue that can´t be solved. We hope that all the efforts made until now in the research field will be rewarded with a greater number of therapeutic proteins produced in plants in the market, for the benefit of society.

5. References

Allen, GC., Spiker, S., & Thompson, WF. (2000). Use of matrix attachment regions (MARs) to minimize transgene silencing. *Plant Molecular Biology*. Vol. 43, pp. (361-376)

Arlen, PA., Falconer, R., Cherukumilli, S., Cole, A., Cole, AM., Oishi, KK., & Daniell, H. (2007). Field production and functional evaluation of chloroplast-derived interferon-α2b. *Plant Biotechnology Journal*, Vol. 5, No. 4, pp. (511-525)

Arnau, J., Lauritzen, C., Petersen, GE., & Pedersen, J. (2006). Current strategies for the use of affinity tags and tag removal for the purification of recombinant proteins. *Protein Expression and Purification*, Vol. 48, pp. (1-13)

Bardor, M., Loutelier-Bourhis, C., Paccalet, T., Cosette, P., Fitchette, A-C., Vézina, L-P., Trépanier, S., Dargis, M., Lemieux, R., Lange, C., Faye, L., & Lerouge, P. (2003). Monoclonal C5-1 antibody produced in transgenic alfalfa plants exhibits a N-glycosylation that is homogenous and suitable for glyco-engineering into human-compatible structures. *Plant Biotechnology Journal*, Vol. 1, pp. (451–462)

Barrera-Saldaña, HA., Robberson, DL., & Saunders, GF. (1982). Transcriptional products of the human placental lactogen gene. *Journal of Biological Chemistry*, Vol. 257, No. 20, pp. (12399-13404)

Benchabane, M., Goulet, C., Rivard, D., Faye, L., Gomord, V., & Michaud, D. (2008). Preventing unintended proteolysis in plant protein biofactories. *Plant Biotechnology Journal*, Vol. 6, pp. (633-648)

Biemelt, S., Sonnewald, U., Galmbacher, P., Willmitzer, L., & Müller, M. (2003). Production of human papillomavirus type 16 virus-like particles in transgenic plants. *Journal of Virology*, Vol. 77, No. 17, pp. (9211- 9220)

Bock, R., & Khan, MS. (2004). Taming plastids for a green future. *TRENDS in Biotechnology*, Vol. 22, No. 6, (June 2004), pp. (311-318)

Bohem, R. (2007). Bioproduction of therapeutic proteins in the 21st century and the role of plants and plant cells as production platforms. *Annals of the New York Academy of Sciences*, Vol. 1102, pp. (121-134)

Brelje, TC., Scharp, DW., Lacy, PE., Ogren, L., Talamantes, F., Robertson, M., Friesen, HG., & Sorenson, RL. (1993). Effect of homologous placental lactogens, prolactins, and growth hormones on islet β-cell division and insulin secretion in rat, mouse, and human islets: implication for placental lactogen regulation of islet function during pregnancy. *Endocrinology*, Vol. 132, pp. (879-887)

Castañón, S., Martín-Alonso, JM., Marín, MS., Boga, JA., Alonso, P., Parra, F., & Ordás, RJ. (2002). The effect of the promoter on expression of VP60 gene from rabbit hemorraghic disease virus in potato plants. *Plant Science*, Vol. 162, pp. (87-95)

Chen, M., Liu, X., Wang, Z., Song, J., Qi, Q., & Wang, PG. (2005). Modification of plant N-glycans processing: the future of producing therapeutic protein by transgenic plants. *Medicinal Research Reviews*, Wiley Periodicals, Inc, retrieved from www.insterscience.wiley.com.

Commandeur, U., Twyman, RM., & Fischer, R. (2003). The biosafety of molecular farming in plants. *AgBiotechNet*, Vol. 5, pp. (1-9)

Conley, AC., Jevnikar, AM., Menassa, R., & Brandle, JE. (2010). Temporal and spatial distribution of erythropoietin in transgenic tobacco plants. *Transgenic Research*, Vol. 19, No. 2, pp. (291-298)

Corrado, G., & Karali, M. (2009). Inducible gene expression systems and plant biotechnology. *Biotechnology Advances*, Vol. 27, pp. (733-743)

Cox, KM., Sterling, JD., Regan, JT., Gasdaska, JR., Frantz, KK., Peele, CG., Black, A., Passmore, D., Moldovan-Loomis, C., Srinivasan, M., Cuison, S., Cardarelli, PM., & Dickey, LF. (2006). Glycan optimization of a human monoclonal antibody in   the aquatic plant *Lemna minor*. *Nature Biotechnology*, Vol. 24, No. 12, (November 2006), pp. (1591-1597)

Daniell, H., Streatfield, SJ., & Wycoff, K. (2001). Medical molecular farming: production of antibodies, biopharmaceuticals and edible vaccines in plants. *Trends in Plant Science*, Vol. 6, pp. (219-226)

De Carvalho, F., Gheysen, G., Kushnir, S., Van Montagu, M., Inzé, D., & Castresana, C. (1992). Suppression of β-1,3-glucanase transgene expression in homozygous plants. *EMBO Journal*, Vol. 11, No. 7, pp. (2595-2602)

Decker, EL., & Reski, R. (2007). Moss bioreactors producing improved biopharmaceuticals. *Current Opinion in Biotechnology*, Vol. 18, pp. (393–398)

Desai, PN., Shrivastava, N., & Padh, H. (2010). Production of heterologous proteins in plants: strategies for optimal expression. *Biotechnology Advances*, Vol. 28, pp. (427-435)

Dietze J., Blau A. & Willmitzer L. (1995) *Agrobacterium*-mediated transformation of potato (*Solanum tuberosum*), In: *Gene Transfer to Plants XXII*, Potrykus, I & G. Spangenberg, G, pp. (24-29), Springer-Verlag, Berlin

Doran, PM. (2000). Foreign protein production in plant tissue cultures. *Current Opinion in Biotechnology*, Vol. 11, pp. (199-204)

Doran, PM. (2006). Foreign protein degradation and instability in plants and plant tissue cultures. *Trends in Biotechnology*, Vol. 24, No. 9, pp. (426-432)

Engelhard, M. (2007). Pharming-an introduction. In: *Pharming, a new branch of biotechnology*, Engelhard, M., Hagen, K., and Thiele, F. (Eds), pp. (13-31), Europäische Akademie, ISSN 1435-487 X, Germany

Farnsworth, N., Akerele, O., Bingel, A.S., Soejarto, DD., & Guo, Z. (1985). Medicinal plants in therapy. *Bulletin of the World Health Organization*, Vol.63, No. 6, pp. (965-981)

Fiedler, U., Phillips, J., Artsaenko, O., & Conrad, U. (1997). Optimization of scFv antibody production in transgenic plants. *Immunotechnology*, Vol. 3, pp. (205-216)

Fischer, R., Emans, N., Twyman, RM., & Schillberg, S. (2004). Molecular Farming in Plants: Technology Platforms. In: *Encyclopedia of Plant and Crop Science*, Dekker, M. (ed), pp. (753-756), Marcel Dekker Inc, New York

Floss, DM., Schallau, K., Rose-John, S., Conrad, U., & Scheller, J. (2009). Elastin-like polypeptides revolutionize recombinant protein expression and their biomedical application. *Trends in Biotechnology*, Vol. 28, No. 1, (November 2009), pp. (37-45)

Fojtova, M., Van Houdt, H., Depicker, A., & Kovarik, A. (2003). Epigenetic switch from posttranscriptional to transcriptional silencing is correlated with promoter hypermethylation. *Plant Physiology*, Vol. 133, pp. (1240-1250)

Franconi, R., Demurtas, OC., & Massa, S. (2010). Plant-derived vaccines and other therapeutics produced in contained systems. *Expert Review Vaccines*, Vol. 9, No. 8, pp. (877-892), ISSN 1476-0584

Fujinaka, Y., Takane, K., Yamashita, H., & Vasavada, RC. (2007). Lactogens promote beta cell survival through JAK2/STAT5 activation and BCL-X_L upregulation. *Journal of Biological Chemistry*, Vol. 282, pp. (30707-30717)

Gerlach, JQ., Kilcoyne, M., McKeown, P., Spillane, C., & Joshi, L. (2010). Plant-produced biopharmaceuticals. In: *Transgenic crop plants*, Kole, C et al. (eds.), pp. (269-299), Springer-Verlag, ISBN: 978-3-642-04811-1, Heidelberg, Germany

Gerngross, TU. (2004). Advances in the production of human therapeutic proteins in yeast and filamentous fungi. *Nature Biotechnology*, Vol. 22, No. 11 (November 2004), pp. (1409-1414)

Gils, M., Kandzia, R., Marillonet, S., Klimyuk, V., & Gleba, Y. (2005). High-yield production of authentic human growth hormone using a plant virus-based expression system. *Plant Biotechnology Journal*, Vol. 3, pp. (613-620)

Gils, M., Marillonnet, S., Werner, S., Grützner, R., Giritch, A., Engler, C., Schachschneider, R., Klimyuk, V., & Gleba, Y. (2008). A novel hybrid seed system for plants. *Plant Biotechnology Journal*, Vol. 6, pp. (226-235)

Gomord, V., & Faye, L. (2004). Posttranslational modification of therapeutic proteins in plants. *Current Opinion in Plant Biology*, Vol. 7, pp. (171–181)

Gomord, V., Sourrouille, C., Fitchette, AC., Bardor, M., Pagny, S., Lerouge, P., & Faye, L. (2004). Production and glycosylation of plant-made pharmaceuticals: the antibodies as a challenge. *Plant Biotechnology Journal*, Vol. 2, pp. (83–100)

Hellwig, S., Drossard, J., Twyman, RM., & Fischer, R. (2004). Plant cell cultures for the production of recombinant proteins. *Nature Biotechnology*, Vol. 22, No. 11, pp. (1415-1422)

Huang, T-K., & McDonald, KA. (2009). Bioreactor engineering for recombinant protein production in plant cell suspension cultures. *Biochemical Engineering Journal*, Vol. 45, pp. (168–184)

Hüns, M., Neuman, K., Hausmann, T., Ziegler, K., Kleme, F., Kahmann, U., Staiger, D., Lockau, W., Pistorius, EK., & Broer, I. (2008). Plastid targeting strategies for cyanophycin synthetase to achieve high-level polymer accumulation in *Nicotiana tabacum*. Plant Biotechnology Journal, Vol. 6, pp. (321-336)

Hunter, CV., Tiley, LS., & Sang, HM. (2005). Developments in transgenic technology: applications for medicine. *Trends in Molecular Medicine*, Vol. 11, No. 6, pp. (293-298)

Hyunjong, B., Lee, D-S., & Hwang, I. (2006). Dual targeting of xylanase to chloroplasts and peroxisomes as a mean to increase protein accumulation in plant cells. *Journal of Experimental Botany*, Vol. 57, No. 1, pp. (161-169)

Jamal, A., Ko, K., Kim, H-S., Choo, Y-K.., Joung, H., & Ko, K. (2009). Role of genetic factors and environmental conditions in recombinant protein production for molecular farming. *Biotechnology Advances*, Vol. 27, pp. (914-923)

James, E., & Lee, JM. (2001). The production of foreign proteins from genetically modified plant cells. *Advances in Biochemical Engineering/Biotechnology*, Vol. 72, pp. (1-156)

Kadlecêk, P., Tichá, I., Haisel, D., Capkova, V., & Schäfer, C. (2001). Importance of in vitro pretreatment for ex vitro acclimatization and growth. *Plant Science*, Vol. 161, pp. (695-701)

Kamenarova, K., Abumhadi, N., Gecheff, K., & Atanassov, A.(2005). Molecular faming in plants: An approach of agricultural biotechnology. *Journal of Cell and Molecular Biology*, Vol. 4, pp. (77-86)

Kay, R., Chan, A., Daly, M., & McPherson, J. (1987). Duplication of CaMV 35S promoter sequences creates a strong enhancer for plant genes. *Science*, Vol. 236, pp. (1299-1302)

Kermode, AR. (2006). Plants as factories for production of biopharmaceutical and bioindustrial proteins: lessons from cell biology. *Canadian Journal of Botany*, Vol. 84, pp. (679-694)

Kim, T-G., Lee, H-J., Jang, Y-S., Shin, Y-J., Kwon, T-H., & Yang, M-S. (2008). Co-expression of proteinase inhibitor enhances recombinant human granulocyte–macrophage colony stimulating factor production in transgenic rice cell suspension culture. *Protein Expression and Purification*, Vol. 61, pp. (117–121)

Kim, H-S., Euym, J-W., Kim, M-S., Lee, B-C., Mook-Jung, I., Jeon, J-H., & Joung, H. (2003). Expression of human β-amyloid peptide in transgenic potato. *Plant Science*, Vol. 165, pp. (1445-1451)

Knäblein, J. (2005). Plant-based expression of biopharmaceuticals, In: *Encyclopedia of Molecular Cell Biology and Molecular Medicine*, R. A. Meyers (Ed), 385-410, Wiley-VCH Verlag GmbH & Co., ISBN 3-527-30552-1, Berlin, Germany

Koprivova, A., Stemmer, C., Altmann, F., Hoffmann, A., Kopriva, S., Gorr, G., Reski, R., &
     Decker, EL. (2004). Targeted knockouts of *Physcomitrella* lacking plant-specific
     immunogenic N-glycans. *Plant Biotechnology Journal*, Vol. 2, pp. (517–523)
Kusnadi, AR., Nikolov, ZL., & Howard, JA. (1997). Production of recombinant proteins in
     transgenic plants: practical considerations. *Biotechnology and Bioengineering*, Vol. 56,
     pp. (473-482)
Kwon, T-H., Kim, Y-S., Lee, J-H., & Yang, M-S. (2003). Production and secretion of
     biologically active human granulocyte-macrophage colony stimulating factor in
     transgenic tomato suspension cultures. *Biotechnology Letters*, Vol. 25, pp. (1571-1574)
Lan, PC., Tseng, CF., Lin, MC., & Chang, CA. (2006). Expression and purification of human
     placental lactogen in *Escherichia coli*. *Protein Expression and Purification*, Vol. 46, pp.
     (285-293)
Lee, D., & Natesan, E. (2006). Evaluating genetic containment strategies for transgenic
     plants. *Trends in Biotechnology*, Vol. 24, No. 3, (March 2006), pp. (109-114)
Lessard, PA., Kulaveerasingam, H., York, GM., Strong, A., & Sinskey, AJ. (2002).
     Manipulating gene expression for the metabolic engineering of plants. *Metabolic
     Engineering*, Vol. 4, pp. (67-79)
Liénard, D.; Sourrouille, C.; Gomord, V. & Faye, L. (2007). Pharming and transgenic plants.
     Biotechnology Annual Review, Vol.13, pp. (115-147), ISSN 1387-2656
Ma, JKC., Chikwamba, R., Sparrow, P., Fischer, R., Mahoney, R., & Twyman, RM. (2005).
     Plant-derived pharmaceuticals–the road forward. *Trends in Plant Science*, Vol. 10,
     No. 12, (November 2005), pp. (580-585)
Maclean, J., Koekemoer, M., Olivier, AJ., Stewart, D., Hitzeroth, II., Rademacher, T., Fischer,
     R., Williamson, AL., & Rybicki, EP. (2007). Optimization of human papillomavirus
     type 16 (HPV-16) L1 expression in plants: comparison of the suitability of different
     HPV-16 L1 gene variants and different cell-compartment localization. *Journal of
     General Virology*, Vol. 88, pp. (1460-1469)
Maliga, P. (2003). Progress towards commercialization of plastid transformation technology.
     Trends in Biotechnology, Vol. 21, No. 1 (January 2003), pp. (20-28)
Mason, HS., Ball, J-M., Shi, J-J., Jiang, X., Estes, MK., & Arntzen, CJ. (1996) Expression of
     Norwalk virus capsid protein in transgenic tobacco and potato ans its oral
     immunogenicity in mice. *PNAS*, Vol. 93, (May 1996), pp. (5335-5340)
Mason, HS., Haq, TA., Clements, JD., & Arntzen CJ. (1998). Edible vaccine protects mice
     against Escherichia coli heat-labile enterotoxin (LT): potatoes expressing a synthetic
     LT-B gene. *Vaccine*, Vol. 16, No. 13, pp. (1336-1343)
Mayfield, SP., & Franklin, SE. (2005). Expression of human antibodies in eukaryotic micro-
     algae. *Vaccine*, Vol. 23, pp. (1828-1832)
Medina-Bolivar, F., Condori, J., Rimando, AM., Hubstenberger, J., Shelton, K., O´Keefe, SF.,
     Bennett, S., & Dolan, MC. (2007). Production and secretion of resveratrol in hairy
     root cultures of peanut. *Phytochemistry*, Vol. 68, pp. (1992-2003)
Menassa, R., Kennette, W., Nguyen, V., Rymerson, R., Jevnikar, A., & Brandle, J. (2004).
     Subcellular targeting of human interleukin-10 in plants. *Journal of Biotechnology*,
     Vol. 108, pp. (179-183)
Min, SR., Woo, JW., Jeong, WJ., Han, SK., Lee, YB., & Liu, JR. (2006). Production of human
     lactoferrin in transgenic cell suspension cultures of sweet potato. *Biologia Plantarum*,
     Vol. 50, No. 1, pp. (131-134)

Molina, A., Hervás-Stubbs, S., Daniell, H., Mingo-Castel, AM., & Veramendi, J. (2004). High-yield expression of a viral peptide animal vaccine in transgenic tobacco chloroplasts. *Plant Biotechnology Journal*, Vol. 2, pp. (141-153)

Murashige, T., & Skoog, F. (1962). A revised medium for rapid growth and bioassay with tobacco tissue cultures. *Physiologia Plantarum*, Vol. 15, pp. (473-497)

Nagata, T., & Kumagai, F. (1999). Plant cell biology through the window of the highly synchronized tobacco BY-2 cell line. *Methods in Cell Science*, Vol. 21, pp. (123-127)

Obembe, O., Popoola, JO., Leelavathi, S., & Reddy, SV. (2010). Advances in plant molecular farming. *Biotechnology Advances*, Vol. 29, No. 2, pp. (210-222)

Peterson, RKD., & Arntzen, CJ. (2004). On risk and plant-based biopharmaceuticals. *Trends in Biotechnology*, Vol. 22, No. 2, (February 2004), pp. (64-66)

Potvin, G., & Zhang, Z. (2010). Strategies for high-level recombinant protein expression in transgenic microalgae: a review. *Biotechnology Advances*, Vol. 28, pp. (910-918)

Rai, M., & Padh, H. (2001). Expression systems for production of heterologous proteins. *Current Science*, Vol. 80, No. 9 (May 2001), pp. (1121-1128)

Rasala, BA., & Mayfield, SP. (2011). The microalga *Chlamydomonas reinhardtii* as a platform for the production of human protein therapeutics. *Bioengineered Bugs*, Vol. 2, No. 1 (January/February 2011), pp. (50-54)

Rates, S.M.K. (2001). Plants as source of drugs. *Toxicon*, Vol. 39, pp. (603-613)

Ritcher, LJ., Thanavala, Y., Arntzen, CJ., & Mason, HS. (2000). Production of hepatitis B surface antigen in transgenic plants for oral immunization. *Nature Biotechnology*, Vol. (18), pp. (1167-1171)

Saint-Jore-Dupas C, Faye L, & Gomord V (2007) From planta to pharma with glycosylation in the toolbox. *Trends in Biotechnology*, Vol. 25, No. 7, pp. (317-323)

Satoh, J., Kato, K., & Shinmyo, A. (2004). The 5'-untranslated region of the tobacco *alcohol dehydrogenase* gene functions as an effective translational enhancer in plants. *Journal of Bioscience and Bioengineering*, Vol. 98, No. 1, pp. (1-8)

Schiermeyer, A., Schinkel, H., Apel, S., Fischer, R., & Schillber, S. (2005). Production of *Desmodus rotundus* salivary plasminogen activator α1 (DSPAα1) in tobacco is hampered by proteolysis. *Biotechnology and Bioengineering*, Vol. 89, No. 7, (March 2005), pp. (848-858)

Schillberg, S., & Twyman, RM. (2007). Pharma-Planta: recombinant pharmaceuticals from plants for human health. In: *Pharming, a new branch of biotechnology*, Engelhard, M., Hagen, K., and Thiele, F. (Eds), pp. (13-31), Europäische Akademie, ISSN 1435-487 X, Germany

Schouten, A., Roosien, J., van Engelen, FA., de Jong, GAM., Borst-Vrenssen, AWM., Zilverentant, JF., Bosch, D., Stiekema, WJ., Gommers, FJ., Schots A., & Bakker, J. (1996). The C-terminal KDEL sequence increases the expression level of a single-chain antibody designed to be targeted to both the cytosol and the secretory pathway in transgenic tobacco. *Plant Molecular Biology*, Vol. 30, pp. (781-793)

Shaaltiel, Y., Bartfeld, D., hashmueli, S., Baum, G., Brill-Almon, E., Galili, G., Dym, O., Boldin-Adamsky, SA., Silman, I., Sussman, JL., Futerman, AH., & Avizer, D. (2007). Production of glucocerebrosidase with terminal mannose glycans for enzyme replacement therapy of Gaucher's disease using a plant cell system. *Plant Biotechnology Journal*, Vol. 5, pp. (579-590)

Sharp, JM., & Doran, PM. (2001). Strategies for enhancing monoclonal antibody accumulation in plant cell and organ cultures. *Biotechnology Progress*, Vol. 17, No. 6, pp. (979-992)

Sijmons, PC., Dekker, BMM., Schrammeijer, B., Verwoerd, TC., van den Elzen, PJM., & Hoekema, A. (1990). Production of correctly processed human serum albumin in transgenic plants. *Biotechnology*, Vol. 8, (March 1990), pp. (217-221)

Sivakumar, G. (2006). Bioreactor technology: a novel industrial tool for high-tech production of bioactive molecules and biopharmaceuticals from plant roots. *Biotechnology Journal*, Vol. 1, pp. (1419-1427)

Sparrow, PAC., Irwin ,JA., Dale, PJ., Twyman, RM., & Ma, JKC. (2007). Pharma-Planta: road testing the developing regulatory guidelines for plant-made pharmaceuticals. *Transgenic Research*, Vol. 16, No. 2, pp. (147-161)

Specht, E., Miyake-Stoner, S., & Mayfield, S. (2010). Micro-algae come of age as a platform for recombinant protein production. *Biotechnology Letters*, Vol. 32, pp. (1373-1383)

Spök, A., Twyman, RM., Fischer, R., Ma, JKC., & Sparrow, PAC. (2008). Evolution of a regulatory framework for pharmaceuticals derived from genetically modified plants. *Trends in Biotechnology*, Vol. 26, pp. (506-517)

Stoger, E., Ma, JKC., Fischer, R., & Christou, P. (2005). Sowing the seeds of success: pharmaceutical proteins from plants. *Current Opinion in Biotechnology*, Vol. 16, pp. (167-173)

Su, WW. (2006). Bioreactor engineering for recombinant protein production using plant cell suspension culture. In: *Plant Tissue Culture Engineering*, Gupta, SD., and Ibaraki, Y ,(Eds), pp 135–159, Springer, ISBN 978-1-4020-3694-1, Netherlands

Svab, Z., & Maliga, P. (2007). Exceptional transmission of plastids and mitochondria from the transplastomic pollen parent and its impact on transgene containment. *PNAS*, Vol. 104, No. 17, (april 2007), pp. (7003–7008)

Torres, E., Vaquero, C., Nicholson, L., Sack, M., Stöger, E., Drossard, J., Christou, P., Fischer, R., & Perrin, Y. (1999). Rice cell culture as an alternative production system for functional diagnostic and therapeutic antibodies. *Transgenic Research*, Vol. 8, pp. (441-449)

Tremblay, R., Wang, D., Jevnikar, AM., & Ma, S. (2010). Tobacco, a highly efficient green bioreactor for production of therapeutic proteins. *Biotechnology Advances*, Vol. 28, pp. (214–221)

Twyman, RM., Schillberg, S., & Fischer, R. (2005). Transgenic plants in the biopharmaceutical market. *Expert Opinion. Emerging Drugs*, Vol. 10, No.1, ISSN 1472-8214

Twyman, RM., Stoger, E., Schillberg, S., Christou, P., & Fischer, R. (2003). Molecular farming in plants: host systems and expression technology. *Trends in Biotechnology*, Vol. 21, No. 12 (December 2003), pp. (570-578)

Ulker, B., Allen, GC., Thompson, WF., Spiker, S., & Weissinger, AK. (1999). A tobacco matrix attachment region reduces the loss of transgene expression in the progeny of transgenic tobacco plants. *Plant Journal*, Vol. 18, pp. (253-263)

Urreta, I., Oyanguren, I., & Castañón, S. (2010). Tobacco as biofactory for biologically active hPL production: a human hormone with potential applications in type-1 diabetes. *Transgenic Research*, Vol. 20, No. 4, pp. (721-733)

Vasavada, RC., García-Ocaña, A., Zawalich, WS., Sorenson, RL., Dann, P., Syed, M., Ogren, L., Talamantes, F., & Stewart, AF. (2000). Targeted expression of placental lactogen in the   beta cells of transgenic mice results in beta cell proliferation, islet mass augmentation, and hypoglycemia. *Journal of Biological Chemistry*, Vol. 275, pp. (15399-15406)

Vitale, A., & Pedrazzini, E. (2005). Recombinant pharmaceuticals from plants: the endomembrane system as bioreactor. *Molecular Interventions*, Vol. 5, No. 4, pp. (216-225)

Walker, TL., Purton, S., Becker, DK., & Collet, C. (2005) Microalgae as bioreactors. *Plant   Cell Reports*, Vol. 24, pp. (629–641)

Walsh, G. (2005a). Current Status of Biopharmaceuticals: Approved Products and Trends in Approvals, In: *Modern Biopharmaceuticals*. Knäblein, J, pp. (1-34), WILEY-VCH Verlag GmbH & Co. KGaA, ISBN: 3-527-31184-X, Weinheim

Walsh, G. (2005b). Biopharmaceuticals: recent approvals and likely directions. *Trends in Biotechnology*, Vol. 23, No. 11 (November 2005), pp. (553-558)

Walsh, G. (2010). Biopharmaceutical benchmarks 2010. *Nature Biotechnology*, Vol. 28, No. 9 (September 2010), pp. (917-924)

Wang, M-L., Goldstein, C., Su, W., Moore, PH., & Albert, HH. (2005). Production of biologically active GM-CSF in sugarcane: a secure biofactory. *Transgenic Research*, Vol. 14, pp. (167-178)

Wang, X., Brandsma, M., Tremblay, R., Maxwell, D., Jevnikar, AM., Huner, N., & Ma, S. (2008). A novel expression platform for the production of diabetes-associated autoantigen human glutamic acid decarboxylase (hGAD65). *BMC Biotechnology*, Vol. 8, No. 87, (November 2008)

Weathers, PJ., Towler, MJ., & Xu, J. (2010). Bench to batch: advances in plant cell culture for producing useful products. *Applied Microbiology and Biotechnology*, Vol. 85, pp. (1339-1351)

Wirth, S., Segretin, MA., Mentaberry, A., & Bravo-Almonacid., F. (2006). Accumulation of hEGF and hEGF–fusion proteins in chloroplast-transformed tobacco plants is higher in the dark than in the light. *Journal of Biotechnology*, Vol. 125, pp. (159–172)

Xu, J., Okada, S., Tan, L., Goodrum, KJ., Kopchick, JJ., & Kieliszewski, MJ. (2010). Human growth hormone expressed in tobacco cells as an arabinogalactan-protein fusion glycoprotein has a prolonged serum life. *Transgenic Research*, Vol. 19, No. 5, pp. (849-867)

Xu, J., Ge, X., & Dolan, MC. (2011). Towards high-yield production of pharmaceutical proteins with plant cell suspension cultures. *Biotechnology Advances*, Vol. 29, No. 3, (May-June, 2011), pp. (278-299)

Zhou, J-Y., Wu, J-X., Cheng, L-Q., Zheng, X-J., Gong, H., Shang, S-B., & Zhou, E-M. (2003). Expression of immunogenic S1 glycoprotein of infectious bronchitis virus in transgenic potatoes. *Journal of Virology*, Vol. 77, No.16, (August 2003), pp. (9090-9093)

5

Trichome Specific Expression: Promoters and Their Applications

Alain Tissier

Department of Cell and Metabolic Biology, Leibniz Institute of Plant Biochemistry,
Weinberg,
Germany

1. Introduction

As often reminded to the readers in articles or reviews which deal with plant adaptation to their environment, higher plants are sessile organisms, a life habit which does not allow them to escape danger or to move to avoid adverse conditions. This environmental pressure has led to a myriad of adaptations, which are reflected in the vast diversity of plant habitats, morphologies, life cycles and physiological adaptations among others. The surface of the aerial parts of plants is a major interaction domain between the plant and its environment and as such is the site of many adaptations, be they chemical or anatomical. Among those adaptations, the leaf hairs or trichomes, which cover the surface of a large number of plant species, play a prominent role. Plant trichomes constitute a world of their own, so great is their diversity. In a review published in 1978 and entitled "A glossary of plant hair terminology", Payne compiles a comprehensive list of more than 490 terms used to describe trichome morphology (Payne, 1978). Despite this extensive diversity, two major classes of trichome may be distinguished on the basis of their capacity to produce and secrete or store significant quantities of secondary metabolites, namely glandular or non-glandular. Non-glandular trichomes, or leaf hairs, are poorly metabolically active and provide protection mainly through physical means, for example by restricting access to insects, but also by preventing water losses, or protecting against UV radiation. *Arabidopsis thaliana* has been a model for the study of non-glandular trichome development and many genes involved in non-glandular trichome initiation and development could be identified and characterized (Uhrig and Hulskamp, 2010). The metabolic activity of these non-glandular trichomes is however fairly limited and offers little potential for metabolic engineering. A particular class of hairs is the fibers which are present in various species. Cotton seed trichomes are the most economically important since they are the basis of the cotton fiber, but other species such as cottonwood also have fiber hairs. Glandular trichomes are present in many different plant families and can also be divided in two main classes. The capitate trichomes typically have 1 to 10 glandular cells located at the tip of the trichome stalk, and the secretion is directly exuded from the top cells. The secreted material is in general fairly viscous, and in many cases it makes the leaves sticky. Those trichomes are encountered for example in the Solanaceae (tobacco, tomato, potato, etc.) and in some Lamiaceae species (e.g. *Salvia*). Peltate trichomes have the capacity to synthesize and store volatile compounds (mono- and sesquiterpenes, phenylpropenes) in a subcuticular cavity. Typical representative examples are those from mint and other Lamiaceae, which are valued for the essential oil produced in their trichomes. In both cases, the massive metabolic fluxes that take place in the secretory cells may lead to the accumulation of metabolites which represent up to 10-15% of the leaf dry weight (Wagner et al., 2004). These cells can thus be considered like true cell factories and therefore constitute attractive targets for metabolic engineering (Schilmiller et al., 2008).

1.1 Why trichome specific promoters?

Whether they are cotton fibers or glandular trichomes producing essential oils or resins, the availability of genes and promoters which are specifically expressed in those structures provides material both for more in-depth studies of trichome specific processes and for high precision engineering of trichome traits. A number of genes which are highly expressed in trichomes may also be expressed in other organs because they are involved in similar processes there. The promoters from these genes are not ideal for the study of trichome specific processes for obvious reasons. Using these promoters will lead to expression outside of the trichomes and may lead to undesirable effects because of the toxicity of the compounds produced. A trichome specific promoter may be used in several ways to further investigate trichome processes. One is to search for transcription factors by one-hybrid screening or other related methods. Unbiased search for upstream regulators may also be achieved in mutant screens in plants expressing promoter:reporter gene fusions. Although not necessarily practical in the species of interest (for example in mint which is a sterile polyploidy species), a convenient host with conserved features but which is more amenable to transformation and screening, may be chosen for this purpose.

Another major motivation to isolate and characterize trichome specific promoters is genetic engineering, in particular for the expression of metabolic pathway genes. When expressed under a strong ubiquitous promoter, like the Cauliflower Mosaic Virus (CaMV) 35S, perturbation of metabolic pathways in the whole plant may have deleterious consequences on plant development and physiology. The trichomes, as a distinct entity with restricted communication to the rest of plant, represent therefore a particularly interesting target for metabolic engineering.

Besides metabolic engineering, the availability of trichome specific regulators may help to modify trichome related traits. For example, modulating the expression of transcription factors specifically controlling trichome differentiation and/or development could lead to an increase in trichome density, an improvement of the productivity of trichome-based secretions (e.g. essential oils) or a boost in trichome-mediated resistance to insect pests or other pathogens.

2. Cotton

2.1 Genomics of cotton fibers

Cotton fibers are specialized single-celled hairs which develop on ovules. The cotton hairs are among the longest plant cells reported and are coated with cellulose fibers which confer its value to the cotton crop. Because cotton hairs are single-celled, it has been proposed that their development is controlled by similar gene networks as those of Arabidopsis leaf trichomes, which are also single-celled but branched. It should be noted however, that Arabidopsis seeds do not have trichomes and thus cannot be considered as an ideal surrogate model to evaluate the specificity of expression of cotton fiber genes. The development of seed trichomes is a synchronized process with several easily distinguishable phases. These have been well documented in previous reviews and will be briefly summarized here. The initiation of fiber cells takes place early on at the onset of anthesis, which is conveniently used as the reference time point expressed in days post anthesis (DPA) (Lee *et al.*, 2007). Already after 2 DPA, the fibers start elongating, a process which lasts until 20 DPA. This is followed by secondary wall biosynthesis until 45-50 DPA and concluded by the maturation phase. The synchronized process has allowed the preparation of RNA from these different phases. Initially, fiber specific genes were isolated by differential screening of cDNA library. This led to the successful identification of several genes with strong and specific expression in fibers, including E6, genes encoding Lipid Transfer Proteins (LTPs), a Proline Rich Protein and other genes with no obvious sequence similarity (John and Crow, 1992; Ma *et al.*, 1995; Orford and Timmis, 1995; Rinehart *et al.*, 1996; Orford and Timmis, 1997; Orford and Timmis, 1998; Orford *et al.*, 1999). Already, Northern or RT-PCR analysis showed that genes can be expressed during distinct phases of development of the fiber cells or throughout the life of these cells. This is relevant since the promoters from these genes should allow to direct the expression of transgenes during given stages of development of the fibers, which may have important practical consequences depending on the engineering objective. These early studies were followed by genomics approaches, including Expressed Sequence Tag (EST) library sequencing and microarray hybridization. In particular, EST libraries corresponding to various stages of development were produced and these provide invaluable resources for the identification of fiber specific genes (Li *et al.*, 2002a; Arpat *et al.*, 2004; Udall *et al.*, 2006; Yang *et al.*, 2006). As genes from these EST collections start being characterized, more information has become available on the pattern of expression and the importance of some transcription factors in fiber development (Lee *et al.*, 2007). For some of the genes, the promoters have been cloned and characterized by transgenesis or transient assays. Because cotton transformation is a lengthy process, alternative hosts have been used to characterize cotton promoter:GUS fusions. In most cases, these are either *Arabidopsis thaliana* or tobacco (*Nicotiana tabacum*). These hosts are far from ideal when it comes to characterize seed fiber specific expression because they are both devoid of seed trichomes. Arabidopsis is perhaps a little better because its trichomes are single-celled, like those of cotton, whereas those of tobacco are typically multicellular. There is, in addition, evidence that single celled trichomes from Arabidopsis, which, like cotton, belongs to the Rosids, and multicellular trichomes of the Solanaceae or other Asterids (Antirrhinum) are under the control of distinct regulatory network (Serna and Martin, 2006). A list of available cotton fiber promoters is provided in Table 1. This list is probably not exhaustive, but contains already 28 promoters, underscoring the high interest in characterizing such promoters. The expression range, expressed in DPA was compiled, and illustrates the diversity of promoters available, from the differentiation stage to the late secondary wall synthesis phase. Thus, targeting engineering to specific phases of fiber development is theoretically possible. It is difficult to compare the strength of these promoters between them, as they were often assessed in independent studies using different methods (Northern, semi-quantitative and quantitative RT-PCR,). Nonetheless, it can be assumed that genes with a function in cell wall biosynthesis, e.g. cellulose synthase, are probably among the most highly expressed.

Gene	Protein description	Expression measured by RT-PCR or Northern	Expression window in cotton fibers (in days)	Expression in other tissues	References
GhE6	hypothetical	Y	15-24	N	(John and Crow, 1992)
GhLTPx_GH3	Lipid transfer protein	Y	5-20	N	(Ma et al., 1995)
FbL2a	Hypothetical	Y	25-45	N	(Rinehart et al., 1996)
pGhEX1	Expansin	Y	6-20	N	(Orford and Timmis, 1998; Harmer et al., 2002)
GhLTP6	Lipid transfer protein	N	10-20	N	(Ma et al., 1995; Hsu et al., 1999)
GhLTP3	Lipid transfer protein	N	5-20	N	(Liu et al., 2000)
GhTUB1	beta-tubulin	Y	0-14	early seedling development (cotyledons, root tips)	(Li et al., 2002b)
GhCTL1-2	Chitinase-like	Y	8-31	xylem, pollen, cells with secondary walls (weak)	(Zhang et al., 2004)
GaRDL1	RD22_like	Y	3-12	N	(Wang et al., 2004)
GhACT1	Actin	Y	4-21	Cotyledons	(Li et al., 2005)
GhDET2	Steroid reductase	Y	3-14	Roots	(Luo et al., 2007)
GhGlcAT1	glucuronosyltransferase	N	NA	NA	(Wu et al., 2007)
Fsltp4	Lipid transfer protein	Y	6-14	N	(Delaney et al., 2007)
GhTUA9	alpha-Tubulin	Y	5-10	N	(Li et al., 2007)
GaHOX1/2	Transcription factor	Y	3-12	N	(Guan et al., 2008)
GaMYB2	Transcription factor	Y	0-9	trichomes in other organs	(Wang et al., 2004; Shangguan et al., 2008)
GhMYB109	Transcription factor	Y	4-8	N	(Suo et al., 2003; Pu et al., 2008)
GhSCFP	Protease	N	2-25	N	(Hou et al., 2008)
GhH6L	Arabinogalactan	Y	3-20	N	(Wu Y, 2009)
GhMYB25	Transcription factor	Y	0-5	trichomes of other tissues, pollen, anthers, root epidermis, root initials	(Machado et al., 2009)

Gene	Protein description	Expression measured by RT-PCR or Northern	Expression window in cotton fibers (in days)	Expression in other tissues	References
GhSUS3	Sucrose synthase	Y	0-5	NA	(Ruan et al., 2009)
GhXTH1	Xyloglucan endotransglycosyl ase/hydrolase	Y	10-25	N	(Michailidis et al., 2009)
GbML1	Transcription factor	Y	-3-8	Petal	(Zhang et al., 2010)
GhRING1	Ubiquitin Ligase	Y	0-20	NA	(Ho et al., 2010)
GhXTH1	Xyloglucan endotransglycosyl ase/hydrolase	Y	10-15	Petal	(Lee et al., 2010)
ADPGp_SSU2	ADP-glucose pyrophosphorylas e	Y	10	meristem, immature stem, roots	(Taliercio, 2011)
GhCesA4	Cellulose synthase	Y	16-24	root vascular tissue	(Wu et al., 2009; Kim et al., 2011)

Table 1. Promoters expressed in cotton fibers. DPA: days post-anthesis. Y: yes; N: no; NA: not available

2.2 Examples of engineering of cotton trichomes

The first attempts at genetic engineering of cotton fibers were performed in the late 1990s, soon after the first specific promoters were identified. The objective was to introduce poly-hydroxybutyrate (PHB) into cotton fibers, via the expression of two genes *phaB* and *phaC* from the bacterium *Alcaligenes eutrophus*, which naturally produces PHB in inclusion bodies. *phaB* encodes the acetoacetyl-CoA reductase and *phaC* the PHB synthase. Expression of both genes in *Arabidopsis thaliana* was previously shown to support *de novo* biosynthesis of PHB in plants for the first time (Poirier *et al.*, 1992). In cotton, this was achieved by expressing *phaB* under the control of the promoters from the fiber specific genes FbL2a or E6, and *phaC* with the FbL2a or 35S promoters. Since the substrate for the PHB synthase does not occur naturally in plants, the expression of *phaC* under 35S should not have deleterious effects on whole plants. The transgenic plants were reported briefly in a first paper (Rinehart *et al.*, 1996) and analyzed in more detail in a second article (John and Keller, 1996). Production of PHB in the lumen of cotton fiber cells could be shown as evidenced by staining, electron microscopy, HPLC and GC-MS. PHB accumulated in the form electron-translucent granules. Quantification of crotonic acid released after hydrolysis indicated levels of up to 3440 µg/g dry fiber in the best lines. The majority of the PHB produced (68.3 %) had a MW above 0.6×10^6 Da, which is similar to PHB produced in bacteria. PHB synthesis peaked at 10 DPA and did not increase nor decrease afterwards, indicating the absence of major PHB degrading activity in cotton fibers. The thermal properties of the transgenic fibers were also assessed and indicated that they had higher heat retention capacity (John and Keller, 1996). However, although promising, those modified properties were apparently not significant enough to warrant commercialization. This was due to the relatively low level of PHB produced (0.34% of fiber weight), which would need to increase several fold to be considered for commercialization.

In a more recent attempt at metabolic engineering, melanin biosynthesis was introduced in cotton fibers (Xu et al., 2007). Dyeing cotton fibers has a heavy imprint on the environment and solutions to reduce its polluting impact are desirable. Naturally colored cotton fibers exist but the choice of colors is limited and the colored cotton varieties have low producing capacity. An alternative is to use biotechnology to engineer colors into cotton fibers. As a proof of concept, Xu and co-workers (2007) expressed two genes, TyrA and ORF438, from Streptomyces antibioticus, which are required and sufficient to synthesize melanin. Both genes were codon optimized for expression in cotton, fused to a vacuolar targeting peptide and cloned under the control of a fiber specific promoter from the Ltp3 gene (Liu et al., 2000). The same construct was used to transform tobacco and cotton. Both in tobacco and in cotton transgenic plants the change in color in the leaf trichomes (tobacco) or in the seed fibers (cotton) was distinctly visible although no dosage of melanin was reported (Xu et al., 2007). In addition to its color, melanin also absorbs UV light and could therefore provide UV-protection properties to cotton fabrics.

3. Tobacco

Tobacco (Nicotiana tabacum) is an allotetraploid species which is grown worldwide for its leaf which is processed and used for various products, from which the most widely sold and consumed are cigarettes. It is well established that regular tobacco smoking is a health-damaging habit with associated increased risks of cancer and cardio-vascular diseases. Health-promoting uses of tobacco could provide alternative revenue sources for tobacco farmers, for example by producing pharmaceutical ingredients in tobacco through genetic engineering. Plant Made Pharmaceuticals (PMPs) have mostly concerned therapeutic proteins, such as antibodies or hormones like insulin. Plants are also known to provide many natural small molecules to the pharmacopeia or as drug leads. These belong to the secondary, or specialized, as they are now sometimes called, classes of metabolites. The huge diversity of these compounds provides a phenomenal reservoir of chemical structures whose biosynthesis pathways are now beginning to be elucidated thanks to the contribution of genomics approaches in plant biochemistry studies. One issue which is frequently raised about plant natural products is the availability of the raw material and the cost associated to extraction and purification of the compound. Pharmaceutical companies will shy away from substances whose supply cannot be safely guaranteed, which is likely to be the case if the chemical is produced in one rare plant of the Amazon forest for example. But the plant does not need to come from tropical forest to be endangered. The story of Taxol is a good example in this respect. Taxol is a diterpenoid extracted from yew tree with potent anticancer activity. Initially, Taxol was extracted from the barks of pacific yew trees (Taxus brevifolia), where it was present in less than 0.01% of the dry matter, with many related taxoids to separate it from, making it an extremely expensive chemical to produce. Chemical synthesis was too complex to be exploited commercially. Since the extraction was destructive, natural populations of Taxus were threatened through commercial exploitation of the trees. Fortunately, a semi-synthetic method starting from a precursor abundant in the twigs, 10-deacetyl-baccatin III, was developed. This allowed a durable and renewable procedure since twigs can be harvested without felling trees.

The presumed progenitors of *N. tabacum* are *N. sylvestris* and *N. tomentosiformis*. All three species have glandular capitate trichomes on their leaf and stem surfaces, with distinct exudate profiles. Cultivated tobacco (*Nicotiana tabacum*) and its wild relatives, *Nicotiana sylvestris* and *N. tomentosiformis* produce diterpenes in large amounts in their glandular capitate trichomes. *N. tomentosiformis* secretes large quantities of labdanoid diterpenes. In *N. tabacum*, these may have two types, either macrocyclic cembranoid or bicyclic labdanoids. The cembranoids are also produced by *N. sylvestris* trichomes and include the cembratrien-diols (α- and β-CBT-diols) and their precursors the cembratrien-ols (α- and β-CBT-ols). Labdanoids include Z-abienol and labdene-diol. Depending on the variety, these diterpenoids may be present in varying amounts and combinations. The terpenoid biosynthesis capacity of tobacco glandular trichomes is massive. In the appropriate conditions, the amount of CBT-diols produced by a *N. sylvestris* leaf may represent up to 10% of the leaf dry weight. Early studies performed on tobacco glandular trichomes concluded that the biosynthesis of the diterpenes takes place in the trichome heads themselves (Keene and Wagner, 1985; Kandra and Wagner, 1988; Guo et al., 1994). This, together with the high productivity of tobacco trichomes makes them an ideal target for terpenoid metabolic engineering.

Terpenes are hydrocarbon molecules whose structure is based on repeated units of isoprene. They are derived from two C5 precursors, isopentenyl diphosphate (IPP) and dimethylallyl diphosphate (DMAPP). IPP units can be sequentially added to DMAPP by isoprenyl transferases, thus leading to the major short chain isoprenyl diphosphates, geranyl diphosphate (GPP- C10), farnesyl diphosphate (FPP – C15) and geranylgeranyl diphosphate (GGPP – C20). These isoprenyl diphosphates are the substrates of terpene synthases, which in many cases make cyclic products. These are the origin of the skeletal diversity of terpenes. In tobacco, the pathway to the cembratrien-diols was elucidated and shown to involve two steps. The first is encoded by a multigene family of diterpene synthases, the cembratrien-ol synthases (CBTS), which altogether account for the mix of the two stereoisomers of CBT-ol (α and β) (Wang and Wagner, 2003; Ennajdaoui et al., 2010). The second step is carried out by a cytochrome P450 mono-oxygenase which hydroxylates the CBT-ols at a specific position (Wang et al., 2001; Wang and Wagner, 2003). Since the biosynthesis of trichome diterpenoids specifically takes place in the glandular cells, one way to get trichome specific promoters is to clone the corresponding genes.

3.1 Tobacco trichome specific promoters

The first tobacco trichome specific promoter identified was that of the *CYP71D16* gene, which encodes the CBT-ol hydroxylase (Wang et al., 2002). The gene was itself identified through subtractive cDNA library construction which was followed by the cloning of the promoter. 1.8 kb of the promoter was sufficient to confer a highly specific expression of the GUS reporter gene to the trichomes. Remarkably, only the glandular cells were stained, highlighting the distinct differentiation status between the glandular cells of the head and the non-glandular cells of trichome stalk.

The CBTS genes provided another set of trichome specific promoters. It was found that the CBT-ol synthase activity is encoded by a family of 3 closely related genes, which arose via recent duplication event. These genes share over 90 % identity at the nucleotide level, including in the promoter regions. One of those promoters (pCBTS2a) was further studied with sequential and internal deletions. This allowed the identification of a positive regulatory region and a negative regulatory region. When the inhibitory region is deleted, expression can be detected in the whole leaf epidermis as well as in patches in roots. On the contrary, when the activating region is deleted, no expression at all can be detected (Ennajdaoui et al., 2010). This indicates that the cell specific expression is the result of the unique combination of a broad activating region with an inhibitory region which restricts expression to the desired cells.

Genes involved in the other tobacco labdanoid pathway have been recently identified and one promoter was also identified and characterized as trichome specific (Tissier, unpublished results).

The availability of several distinct promoters with identical specificity and different strengths should broaden the possibilities for metabolic engineering.

3.2 Strategies and example for tobacco trichome engineering using specific promoters

Since tobacco produces diterpenoids, one logical possibility for metabolic engineering is to use tobacco trichomes for the production of heterologous diterpenoid. The substrate, namely GGPP, should be available in non-limiting quantities and in addition, the glandular cells have a machinery which allows them to excrete hydrophobic compounds like diterpenes. To facilitate the detection of heterologous terpenoids, it may be useful to eliminate or reduce the endogenous diterpenoids. This can be achieved by inactivating the CBTS genes. Because they are members of a multigene family which are most likely located at the same chromosomal locus, the most efficient way to achieve this is to use gene silencing technologies. This was done with an antisense construct under the control of a 35S promoter (Wang and Wagner, 2003). However, more efficient silencing was obtained with intron-hairpin constructs targeting the exon 2 of the CBTS genes, under the control of the CBTS2a gene itself. In this case, the best transgenic lines had almost no CBT-diols detectable (Ennajdaoui et al., 2010). There is also the possibility to exploit natural variation in N. tabacum. During a survey of the metabolic profiles of tobacco leaf exudates we have noticed that some cultivars produce labdanoids and no cembranoids, while others produce cembranoids and no labdanoids. Thus, by crossing these cultivars it is theoretically possible to breed new varieties which produce no diterpenoids at all, but which still have the capacity to produce new ones. Once a diterpene-free background has been established by either of these approaches, heterologous diterpene synthases may be cloned behind trichome specific promoters for targeted expression to the glandular cells. This was successfully done for taxadiene synthase (Rontein et al., 2008) with yields of up to 10 µg/g fresh weight.

3.3 Glandular trichome expression as a gene function discovery tool

Several steps of Taxol biosynthesis have been investigated, including the early oxidations of taxadiene which lead to the synthesis of the important semi-synthesis precursor, 10-DABIII (Croteau et al., 2006). The first of these oxidations was shown to be at the C5 position of the taxadiene core, which is necessary to form the so-called oxetane ring (Hefner et al., 1996). Subsequently, a gene encoding taxadiene 5-α-hydroxylase (T-5-OH) was identified and characterized (Jennewein *et al.*, 2004). In order to reconstitute these early steps of the taxol biosynthesis pathway, both genes were expressed in tobacco under the control of trichome specific promoters. Taxadiene synthase (TS) was cloned downstream of the CBTS2a promoter as described above while the T-5-OH was cloned 3' of the CYP71D16 promoter. Surprisingly, no taxadien-5-α-ol could be identified in the transgenic plants expressing both genes. Instead, a new product, which was later found to be 5(12)-oxa-3(11)-cyclotaxane, derived from a complex rearrangement of the taxadiene core upon oxidation, could be identified (Rontein *et al.*, 2008). The activity of the enzyme was then further proved from a protein expressed in yeast. This shows that the tobacco trichome platform was useful to produce sufficient quantities of the product to be structurally characterized. The initial functional assignment was likely misguided by the presence of small amounts of T-5-OH as a by-product in the enzyme assays. Since T-5-OH was the compound that was looked for, the major product may have been ignored. The novel assignment derived from the initial tobacco expression, was later confirmed by expression in *E. coli*, although in this case significant amounts of T-5-OH could be detected (Ajikumar *et al.*, 2010).

In another example, the function of the genes required for the biosynthesis of Z-abienol in tobacco could be confirmed by expression in *N. sylvestris* trichomes. Z-abienol is a labdane diterpenoid whose biosynthesis was predicted to require two successive enzymatic steps (Guo *et al.*, 1994), first a copalyl-diphosphate synthase like then a kaurene synthase like enzyme (Peters, 2010). Two candidate genes were thus identified and expressed in *N. sylvestris*, which does not produce Z-abienol. The exudate of these transgenic plants contained significant amounts of Z-abienol of up to 100 µg/g FW (Sallaud and Tissier, unpublished results).

The supply of isoprenyl diphosphates IPP and DMAPP in tobacco trichomes should allow also engineering of other terpenoid classes, like mono- or sesquiterpenes. In those cases, the appropriate isoprenyl transferases (i.e. geranyl diphosphate or farnesyl diphosphate synthases) should be expressed in addition to the terpene synthases. This was done for a sesquiterpene synthase from tomato, the santalene and bergamotene synthase, which uses an unusual isoprenyl diphosphate precursor, Z,Z-FPP (Sallaud *et al.*, 2009). Both enzymes are naturally targeted to the plastids, which is where IPP and DMAPP from the methyl-erythritol pathway (MEP) are synthesized. In this case, the sesquiterpenes could not be identified in the leaf exudate, rather in the headspace collected from transgenic plants (Sallaud *et al.*, 2009). This indicates that tobacco trichomes are suitable for the biosynthesis of sesquiterpenes, but not for their storage. For these, and other volatile compounds, such as monoterpenes or phenylpropenes, glandular trichomes with a storage compartment for volatile compounds, such as the peltate trichomes of mint, should be used (see below).

3.4 Other examples of tobacco trichome engineering

As in cotton, the genes for the biosynthesis of the pigment melanin were expressed in tobacco under the control of the cotton trichome promoter LTP3 (Xu *et al.*, 2007). This promoter was previously shown to be active in tobacco trichomes (Liu *et al.*, 2000). Based on the color of trichomes, the presence of melanin could be detected, however no quantification was performed. This, however, shows that glandular trichomes which are normally producing terpenoids may also be used as an engineering platform for other classes of compounds.

4. Mint and other lamiaceae

4.1 Mint

Peppermint (*Mentha x piperita*) is an aromatic plant which is grown worldwide for its essential oil whose distinctive character is imparted by its most well known compound, (-)-menthol. The essential oil of mint, and of many other aromatic plants form the Lamiaceae, is stored in glandular trichomes of the peltate type (Gershenzon *et al.*, 1987; Gershenzon *et al.*, 1989). Peltate trichomes are composed of 8 glandular cells topped by a subcuticular space where the secretion products are stored. When the cuticle is ruptured, by pressing the leaf between the fingers, or by an insect, the volatile compounds are released and may reach their target. It was shown that the peltate trichomes are not just a site of storage, but also that the terpenoids are produced in the peltate glandular cells (Gershenzon *et al.*, 1989). A technique for the purification of intact peltate glands was developed to allow the production of a trichome specific EST library (Gershenzon *et al.*, 1992). This EST library provided sequence information for the characterisation of the (-)-menthol biosynthetic pathway, which was completely elucidated over the years by the research group from Prof. Croteau (Alonso *et al.*, 1992; Gershenzon *et al.*, 1992; Rajaonarivony *et al.*, 1992; Lupien *et al.*, 1999; Turner *et al.*, 1999; Gershenzon *et al.*, 2000; McConkey *et al.*, 2000; Bertea *et al.*, 2001; Wust *et al.*, 2001; Croteau *et al.*, 2005). Thus, peltate trichomes are extremely well adapted for the production and storage of volatile compounds, in particular mono- and sesquiterpenoids. Metabolic engineering of mint trichomes should therefore yield particularly interesting results for these volatile compounds. Mint transformation by *Agrobacterium tumefaciens* was independently reported by several groups (Diemer *et al.*, 1998; Weller *et al.*, 1998). However, although the use of trichome specific promoters was proposed as early as 1999 as a pre-requisite for metabolic engineering in mint (Lange and Croteau, 1999), to our knowledge no characterization of trichome promoters from mint has been published to date. The promoter of the Arabidopsis GL1 transcription factor was shown to be functional in tobacco and peppermint (Gutierrez-Alcala *et al.*, 2005). However, whether the strength of this promoter will be sufficient for metabolic engineering remains to be seen. Nonetheless, the whole menthol pathway from spearmint provides a set of genes with trichome specific expression and the identification of their promoters should not raise major difficulties.

4.2 Basil

Like mint, Basil (*Ocimum basilicum*) is grown for its aromatic properties which are due to volatile compounds produced in similar peltate trichomes. Following the successful approach developed in mint, trichome specific EST libraries from different cultivars of basil, corresponding to distinct chemotypes, were produced. These were used to elucidate the pathways to volatile phenylpropenes and monoterpenes (Gang *et al.*, 2001; Gang *et al.*, 2002a; Gang *et al.*, 2002b; Iijima *et al.*, 2004a; Iijima *et al.*, 2004b). As in mint, the enzymes of the pathway are likely to be highly specific to the peltate glandular cells, and therefore the promoters of the corresponding genes should drive specific expression to these cells. Like mint, basil could prove an interesting host for the metabolic engineering of volatile compounds, with the additional option of the capacity to engineer phenylpropanoid metabolism in accessions which produce phenylpropenes. However, to date no promoters of basil trichome genes have been characterized. One could also assume that promoters from mint should operate in basil, and reciprocally, because of the similarity of their trichomes and the fact that mint and basil both belong to the Lamiaceae.

4.3 Sage

Sage (*Salvia sp.*) is a large genus with a number of species which are grown commercially for the extraction of fragrant or aromatic oils. One of the most important is *Salvia sclarea* (clary sage), a biennial plant which produces both an essential oil rich in linalyl acetate and linalool, and a concrete with high amounts of the labdanoid diterpene sclareol. Sclareol is currently used in the fragrance industry as a synthesis precursor for Ambrox®, a highly valued compound with amber-like fragrance und excellent fixative properties (Decorzant *et al.*, 1987; Martres *et al.*, 1993; Koga *et al.*, 1998; Moulines *et al.*, 2001; Barrero *et al.*, 2004; Moulines *et al.*, 2004). *Salvia sclarea* possesses two types of glandular trichomes, capitate and peltate. The capitate trichomes are likely to produce sclareol, which is secreted onto the surface of the inflorescences while the volatile compounds like linalyl acetate are more likely to be produced in peltate trichomes (Lattoo *et al.*, 2006; Schmiderer *et al.*, 2008). The productivity of sclareol by *Salvia sclarea* is very high, making it an attractive target for metabolic engineering of terpenoids. Recently, massive sequencing of calyx RNA, where peltate and capitate glands are highly abundant, was carried out (Legrand *et al.*, 2010). A number of genes encoding proteins with clear similarities to terpene synthases and enzymes isoprenoid metabolism could be identified, thus providing genes with potentially highly specific pattern of expression, notably restricted to trichome glandular cells. Although transformation and regeneration of transgenic *Salvia sclarea* plants has not been achieved to date, hairy root cultures were established (Kuzma *et al.*, 2006; Kuzma *et al.*, 2008), and transformation of a related species (*Salvia miltiorrhiza*) by *Agrobacterium tumefaciens* could be successfully demonstrated (Yan and Wang, 2007; Lee *et al.*, 2008). These results suggest that trichome specific metabolic engineering of clary sage is technically feasible.

4.4 Lavender

Lavender (*Lavandula angustifolia, L.* x *intermedia* and other species) is a perennial plant grown in the Mediterranean area for its highly fragrant and characteristic essential oil, which is a complex mixture of mono- and sesquiterpenoids. As for other Lamiaceae species discussed above, the essential oil is produced in peltate glandular trichomes located mostly on the inflorescences (Guitton *et al.*, 2010). Here also, genomics approaches have been initiated to better understand the molecular basis of essential oil production. In one study, EST libraries from flowers and leaves were sequenced by the Sanger method to yield a total of 14,000 sequences (Lane *et al.*, 2010), thus providing the foundation to identify trichome specific genes. A recent study also showed that the oil profile changes over the course of flower development correlated with changes in expression of certain terpene synthases, providing important information regarding harvest time (Guitton *et al.*, 2010). In addition, a trichome specific promoter from the linalool synthase of *L. angustifolia* (LaLIS) was recently isolated and characterized (Biswas *et al.*, 2009). Lavender transformation is also well established, having been reported by two independent groups (Mishiba *et al.*, 2000; Nebauer *et al.*, 2000). Attempts were also made at metabolic engineering of essential content by overexpressing 3-hydroxy-3-methylglutaryl CoA reductase (HMGR) and a limonene synthase with a constitutive 35S promoter (Munoz-Bertomeu *et al.*, 2007; Munoz-Bertomeu *et al.*, 2008). Overexpression of HMGR lead to an increase of both monoterpenes and sesquiterpenes, indicating that the cytosolic mevalonate pathway may contribute to both types of terpenes, although monoterpenes are synthesized in the plastids (Munoz-Bertomeu *et al.*, 2007). Overexpression of the spearmint limonene synthase on the other hand led to a strong increase in limonene while not affecting the other constituents of the oil, indicating that the supply of isoprenyl diphosphate precursors is likely not to be limiting in glandular trichomes (Munoz-Bertomeu *et al.*, 2008). These results bode well for the metabolic engineering of lavender trichomes, and no doubt that the availability of specific promoters should allow more precise manipulation of essential production in these species.

5. Tomato

Like tobacco, tomato (*Solanum lycopersicum*) belongs to the family Solanaceae, which is rich in species with trichomes. Wild species of tomatoes, such as *Solanum pennellii*, *S. habrochaites*, and *S. peruvianum* among others, have different trichome types including non-glandular and glandular types. Altogether up to 7 different types could be described, of which 3 main glandular types could be described (Luckwill, 1943)(See Figure 1).

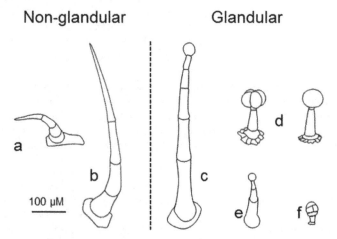

Fig. 1. Trichome types from various tomato species. a. type II non-glandular trichomes from *S. lycopersicum*. b. Type III long hairs (non-glandular). c. Type I long glandular trichomes with single secretory cell at the tip. d. Type VI glandular trichomes. On the left, trichomes from the cultivated tomato, *S. lycopersicum*. On the right, trichomes from the wild species *S. habrochaites*. The type VI trichomes from *S. lycopersicum* have four secretory cells on one plane, which can be easily distinguished from each other. The type VI trichomes from *S. habrochaites* also have four secretory cells, but they are wrapped in a common cuticular envelope, making it look like a single cell from the outside. In addition there is an intercellular space in the middle of these type VI trichomes, where the metabolites are stored. e. type IV trichomes. In some species, like *S. habrochaites*, these trichomes have a single glandular head, while in others like *S. pennellii* they look more like tobacco glandular trichomes with several glandular cells. F. type VII short glandular trichomes.

The type VII are short glandular trichomes with a single stalk cell and a berry-shaped glandular head composed of 7-10 cells. Tobacco also has similar trichomes, and it was shown in tobacco that these trichomes secrete short proline rich proteins, called phylloplanins which have antifungal activities (Shepherd *et al.*, 2005). These trichomes do not appear to secrete small metabolites, and thus seem to be specialized for peptide synthesis. Type I and type IV are capitate trichomes with few or a single glandular cell at the tip. The Type I are extremely long trichomes which can be easily seen with a naked eye, while the type IV trichomes are shorter. Type I trichomes are rare in cultivated tomatoes, while fairly abundant in some wild species, like *S. habrochaites*. Type IV trichomes are absent from cultivated tomatoes, and very abundant in *S. habrochaites*. These trichomes seem to be involved in the synthesis and secretion of secondary metabolites, mostly terpenoids (McDowell *et al.*, 2011). Type VI trichomes are present in both *S. lycopersicum* and *S. habrochaites* but they present distinct morphologies in each species. In *S. lycopersicum*, the four secretory cells are distinctly visible, forming a four-leaf clover shape when viewed from above with a total width of ≈ 60 μM. In *S. habrochaites*, the four secretory cells are encased in an envelope made of cuticule and cell wall materials, so that they appear as a single unit from the outside. The diameter of this ball-shape structure is also about 60 μM, and in contrast to the type VI trichomes from the cultivated tomato, it contains a cavity, most likely of intracellular space resulting from cell wall degradation, between the 4 cells. This storage cavity is reminiscent of the subcuticular space of the peltate trichomes of mint and is likely to contain the secretion products of the glandular cells.

Tomato trichomes have attracted major interest because of their roles in biotic interactions, in particular with arthropods. There are many reports of the roles of trichome secretions in the resistance to insect or arthropod pests (Kennedy, 2003). Most of the resistances to insects are found in wild species, like *S. pennellii* and *S. habrochaites*. For example the white fly *Bemisia tabacci*, which can transfer viruses, can be overcome thanks to glandular trichome secretions (Heinz and Zalom, 1995; Rubinstein and Czosnek, 1997; Snyder *et al.*, 1998; Vendramim *et al.*, 2009). *Tuta absoluta* is another important pest which is recently causing increasing damages to tomato crops. Again sources of resistance have been identified in wild accessions of *S. habrochaites* (Gilardon *et al.*, 2001; Maluf *et al.*, 2010).

The origin of the resistance lies both in the nature and in the quantity of the chemicals secreted. These wild species can be crossed to *S. lycopersicum*, and they can be used to introgress agriculturally relevant traits (disease and abiotic stress resistance, flavor, yield, etc.) into the cultivated tomato genome. However, introgression of complex traits which involve not only biosynthetic pathways, but also regulatory factors controlling may prove difficult and could lead to the introduction of undesirable genes from the wild species which adversely affect yield traits for example, and which may be difficult to eliminate because of the lower level of recombination between wild and cultivated tomato genomes. An alternative is then to introduce the required genes by genetic engineering. To avoid the synthesis of these compounds in the whole plant, which may cause undesirable side effects, trichome specific promoters are required.

In early studies, it was found that polyphenol oxidases are strongly expressed in tomato type VI glandular trichome (Kowalski *et al.*, 1992; Yu *et al.*, 1992; Thipyapong *et al.*, 1997). However, the promoters of this complex multigene family are not specific to trichomes and are also expressed in many other tissues upon stress (Yu *et al.*, 1992; Thipyapong *et al.*, 1997).

5.1 Omics of tomato glandular trichomes

Subsequently, as interest in elucidating the biosynthesis pathways of tomato glandular trichomes increased, trichome specific EST libraries from tomato were produced, in particular from the wild species with abundant trichome secretions. The first libraries were released in 2001 from *S. habrochaites* (accession LA1777))(van Der Hoeven *et al.*, 2000) and *S. pennellii* (accession LA716). They were produced by Sanger sequencing and contained around 2000 sequences each (see Table 2).

Species	accession	Trichome type	Sequencing type	# of ESTs	Reference
S.habrochaites	LA1777	mixed	Sanger	2,656	(van Der Hoeven *et al.*, 2000; Fei *et al.*, 2004)
S. habrochaites	PI126449	mixed	Sanger	5,494	(Fridman *et al.*, 2005)
S. lycopersicum	NA	mixed	Sanger	7,254	(Besser *et al.*, 2009)
S. pennellii	LA716	mixed	Sanger	2,917	(Fei *et al.*, 2004)
S. lycopersicum	LA3475	mixed stems	NGS	278,000	(McDowell *et al.*, 2011)
S. lycopersicum	LA3475	type VII	Sanger	791	(McDowell *et al.*, 2011)
S. lycopersicum	LA3475	type VI	NGS	225,000	(McDowell *et al.*, 2011)
S. lycopersicum	LA3475	type I	Sanger	831	(McDowell *et al.*, 2011)
S. habrochaites	LA1777	mixed leaves	NGS	108,000	(McDowell *et al.*, 2011)
S. habrochaites	LA1777	type I	Sanger	978	(McDowell *et al.*, 2011)
S. habrochaites	LA1777	type IV	Sanger	1,425	(McDowell *et al.*, 2011)
S. habrochaites	LA1777	type VI	NGS	224,000	(McDowell *et al.*, 2011)
S. habrochaites	PI126449	Type VI	Sanger	15,000	(McDowell *et al.*, 2011)
S. pimpinellifolium	LA1589	type VI	NGS	227,000	(McDowell *et al.*, 2011)
S. pennellii	LA0716	type IV	Sanger	1,277	(McDowell *et al.*, 2011)
S. pennellii	LA0716	type VI	Sanger	1,137	(McDowell *et al.*, 2011)
S. pennellii	LA0716	mixed leaves	NGS	275,000	(McDowell *et al.*, 2011)
S- arcanum	LA1708	mixed stems	NGS	415,000	(McDowell *et al.*, 2011)
Total				**1,791,760**	

Table 2. A summary of currently available EST libraries from tomato trichomes. The accession numbers are those according to the Tomato Genetics Resource Center nomenclature (preceded with LA), or from the USDA germplasm collection (preced with PI).

Other similar sequence libraries were produced (Fridman *et al.*, 2005; Slocombe *et al.*, 2008), followed by the recent release of trichome specific libraires from several Solanum species and from distinct trichome types (McDowell *et al.*, 2011). Some of these were sequenced by next generation sequencing technologies, thus affording much larger numbers of EST sequences (up to 278 000 in some cases) (McDowell *et al.*, 2011). These sequence databases have been extremely useful in identifying and characterizing genes for the trichome specific biosynthesis pathways, in particular for terpenes (van Der Hoeven *et al.*, 2000; Sallaud *et al.*, 2009; Schilmiller *et al.*, 2009) and methylketones (Fridman *et al.*, 2005; Ben-Israel *et al.*, 2009; Yu *et al.*, 2010). A summary of the available EST sequences from various tomato species and trichome types is provided in Table 2. With a total of 1 791 760 ESTs, tomato trichomes are probably the trichomes with the best sequence resources currently available.

The availability of the tomato (*S. lycopersicum*) genome sequence makes it possible now to rapidly have access to promoter sequences, although the most interesting promoters will undoubtedly come from the wild species. However, the high sequence similarity between *S. lycopersicum* and *S. pennellii* or *S. habrochaites* should allow the facile identification of the promoters in those wild species.

There is at this stage and to the best of our knowledge, no reported example of metabolic engineering in tomato trichomes. Since tomato is a food crop grown for its fruit, much more has focused on fruit metabolism. There was even a report of taxadiene synthase (TS) expression in tomato fruit under the control of a fruit specific promoter (Kovacs *et al.*, 2007). From a purely metabolic point of view, this makes sense since tomato fruits are rich in carotenoids which derive from the same substrate as diterpenes, namely GGPP. However, the overexpression of TS caused sterility and growth defects which are undesirable side effects. In addition, the presence of potentially toxic secondary compounds in edible vegetables or fruits is a potential source of incidents by contamination of the food supply chain which must be avoided.

6. Artemisia

Artemisia annua, or sweet wormwood, a biennial plant from the Asteraceae family, has attracted attention as the source of an alternative to quinoline drugs for the treatment of malaria. The emergence of foci of resistance to quinine and related drugs in strains of *Plasmodium falciparum* requires the use of durable alternative treatments. Sweet wormwood was long known in Chinese traditional medicine to treat fevers. It was rediscovered in the 1970s for the treatment of malaria. The active ingredient is artemisinin, a sesquiterpene lactone, but semi-synthetic derivatives (Artemeter, Artesunate) have been developed as drugs by the pharmaceutical industry. Artemisia, like many other species from the Asteraceae, produces sesquiterpene lactones in glandular capitate trichomes localized on the leaves, stems and flowers. As with other trichome specific biosynthetic pathways, the elucidation of the first steps of the artemisinin biosynthesis pathways was made possible after sequencing trichome specific cDNA libraries, with the exception of the very first committed step, the sesquiterpene amorphadiene synthase (AaAS). AaAS was initially identified from a leaf cDNA library by similarity to known sesquiterpene synthases from plants (Mercke *et al.*, 2000; Wallaart *et al.*, 2001) and its specific pattern of expression in trichomes was later confirmed (Bertea *et al.*, 2005; Olofsson *et al.*, 2011). A succession of oxidation steps requiring a P450 mono-oxygenase (CYP71AV1) and an aldehyde dehydrogenase leads to artemisinic acid, while the synthesis of de-hydro-artemisinic acid requires the intervention of reductase (DBR2) (Zhang *et al.*, 2008; Liu *et al.*, 2009; Teoh *et al.*, 2009; Wang *et al.*, 2009; Zhang *et al.*, 2009; Weathers *et al.*, 2011). Although much progress has been achieved in the elucidation of the artemisinin pathway, relatively little was done with regards to promoter identification. Although most genes of the pathway are likely to be trichome specific (Liu *et al.*, 2009; Olsson *et al.*, 2009; Wang *et al.*, 2009), only one study reports on the cloning of the *AaAS* promoter and the identification of a WRKY transcription factor (AaWRKY1) which binds to the promoter of *AaAS* (Ma *et al.*, 2009). So far, to the best of our knowledge, no attempt at engineering of *A. annua* trichome metabolism has been reported. Given the importance of this compound as a pharmaceutical ingredient, attempts at metabolic engineering of the artemisinic acid pathway in other plants and in microorganisms will be here briefly reviewed. Artemisinic acid can be used as a precursor for the semi-synthesis of artemisinin and related compounds. In parallel to these pathway elucidation efforts, different approaches were proposed and undertaken to improve the supply of the ingredient. Increasing demand, as well as requirements of reliable quality and supply to maintain stable prices, have spurred the search for either improvement of the available crop plant or transferring the production in heterologous hosts by metabolic engineering.

The first strategy is plant-based with the objective of improving artemisinin production by breeding using existing natural variation or induced mutagenesis. To reach this goal, a high density genetic map based on markers derived from transcriptome deep sequencing was created and used to map QTLs for artemisinin production (Graham et al., 2010). This work showed that next generation sequencing technologies allow the rapid production of dense genetic maps in species where there is little or no prior genetic knowledge.

Another set of approaches is based on the expression of artemisinin biosynthesis genes in heterologous hosts. Reconstitution of the pathway to artemisinic acid was tested in tobacco but using ubiquitous promoters such as CaMV 35S (Zhang et al., 2011). Previous sesquiterpene engineering studies in tobacco had revealed that targeting a FPP synthase together with a sesquiterpene synthase to plastids gave the best results (Wu et al., 2006). With the same strategy and using different combinations of genes, it could be shown that amorphadiene, artemisinic and de-hydroartemisinic alcohol could be produced in the range of $\mu g/g$ FW, but no artemisinic aldehyde or acid could be detected (Zhang et al., 2011). Further analysis indicated that an endogenous reductase in tobacco prevents accumulation of artemisinic aldehyde and acids (Zhang et al., 2011), thus questioning the relevance of tobacco for such metabolic engineering. The use of trichome specific promoters may solve this issue, or perhaps an even better solution would be to use other Asteraceae hosts which are able to accumulate sesquiterpene lactones in large quantities, such as chicory for example. Another explored strategy was the transient expression in N. benthamiana. This system was shown to be quite successful for the transient expression of proteins at very high levels (Marillonnet et al., 2005), however the requirements for successful metabolic engineering are likely to be different. Nonetheless, transient expression of AaAS with CYP71AV1, together with HMGR to increase isoprenyl precursor supply, resulted in the production of artemisinic acid-12-β-diglucoside at levels up to 39.5 mg/kg FW (equivalent 16.6 mg/kg artemisinic acid). This indicated that artemisinic acid is indeed produced to significant levels, and highlights the importance of the host and the tissues targeted for expression. While altogether these results are promising and suggest that metabolic engineering of advanced terpenoid metabolites in plants is feasible, much progress is required to reach levels which will make commercial exploitation a reality. Many combinations of constructs with different promoters, sub-cellular targeting (plastids, cytosol, mitochondria), and hosts will have to be tested to identify the best solutions. However, in plants, even with transient expression systems, this is a highly time consuming tasks.

In comparison, micro-organisms allow a much higher throughput to test a multiplicity of constructs in a short time frame. Highly successful engineering endeavors have been achieved in E. coli and yeast by Keasling and co-workers. Through introduction of mevalonate pathway genes in *E. coli*, production levels of amorphadiene of up to 0.5 g/L could be reached (Martin *et al.*, 2003; Newman *et al.*, 2006). Even artemisinic acid could be produced in *E. coli* after extensive modification of the P450 CYP71AV1 (Chang *et al.*, 2007). However the best results were obtained in yeast, where production levels of up to 100 mg/L could be reached (Ro *et al.*, 2006).

7. Conclusion

Trichomes have been used as a model to study cell differentiation and organ development in *Arabidopsis thaliana*, where the power of molecular genetics and genomics has made possible numerous advances in this area. However, Arabidopsis trichomes offer little opportunities for the development of novel products or applications, essentially because Arabidopsis trichomes are devoid of metabolic or structural properties of interest. However, trichomes play important roles in several crop species, where they are at the origins of important agricultural derived products, like cotton fibers and essential oils and fragrance ingredients of the Lamiaceae. In addition, trichome-borne resistances to insects and microorganisms in plants like tomatoes have attracted interest to restrict the use of pesticides. In those species (cotton, tomatoes, Lamiaceae), extensive EST resources were created and have proved valuable tools to identify and characterize trichome specific genes involved in development or metabolic pathways. Nonetheless, examples of trichome engineering using trichome specific promoters are still scarce and are limited to a handful of cases in cotton and tobacco. It seems that one limitation is to reach levels of productions for the metabolite of interest which are in the same range as those of endogenous metabolites. To reach those levels, it is necessary to understand more about how gene expression in those specialized cells is regulated, so as to be able to design and construct appropriate expression vectors enabling to reach these targets.

8. References

Ajikumar, P.K., Xiao, W.H., Tyo, K.E., Wang, Y., Simeon, F., Leonard, E., Mucha, O., Phon, T.H., Pfeifer, B. and Stephanopoulos, G. (2010). Isoprenoid pathway optimization for Taxol precursor overproduction in Escherichia coli. *Science* 330, 70-4.

Alonso, W.R., Rajaonarivony, J.I., Gershenzon, J. and Croteau, R. (1992). Purification of 4S-limonene synthase, a monoterpene cyclase from the glandular trichomes of peppermint (Mentha x piperita) and spearmint (Mentha spicata). *J. Biol. Chem.* 267, 7582-7.

Arpat, A.B., Waugh, M., Sullivan, J.P., Gonzales, M., Frisch, D., Main, D., Wood, T., Leslie, A., Wing, R.A. and Wilkins, T.A. (2004). Functional genomics of cell elongation in developing cotton fibers. *Plant Mol. Biol.* 54, 911-929.

Barrero, A.F., Alvarez-Manzaneda, E.J., Chahboun, R. and Arteaga, A.F. (2004). Degradation of the side chain of (-)-sclareol: A very short synthesis of nor-Ambreinolide and ambrox. *Synth. Commun.* 34, 3631-3643.

Ben-Israel, I., Yu, G., Austin, M.B., Bhuiyan, N., Auldridge, M., Nguyen, T., Schauvinhold, I., Noel, J.P., Pichersky, E. and Fridman, E. (2009). Multiple Biochemical and Morphological Factors Underlie the Production of Methylketones in Tomato Trichomes. *Plant Physiol.* 151, 1952-1964.

Bertea, C.M., Freije, J.R., van der Woude, H., Verstappen, F.W.A., Perk, L., Marquez, V., De Kraker, J.W., Posthumus, M.A., Jansen, B.J.M., de Groot, A. et al. (2005). Identification of intermediates and enzymes involved in the early steps of artemisinin biosynthesis in Artemisia annua. *Planta Medica* 71, 40-47.

Bertea, C.M., Schalk, M., Karp, F., Maffei, M. and Croteau, R. (2001). Demonstration that menthofuran synthase of mint (Mentha) is a cytochrome P450 monooxygenase: cloning, functional expression, and characterization of the responsible gene. *Arch. Biochem. Biophys.* 390, 279-86.

Besser, K., Harper, A., Welsby, N., Schauvinhold, I., Slocombe, S., Li, Y., Dixon, R.A. and Broun, P. (2009). Divergent regulation of terpenoid metabolism in the trichomes of wild and cultivated tomato species. *Plant Physiol.* 149, 499-514.

Biswas, K.K., Foster, A.J., Aung, T. and Mahmoud, S.S. (2009). Essential oil production: relationship with abundance of glandular trichomes in aerial surface of plants. *Acta Physiol Plant* 31, 13-19.

Chang, M.C.Y., Eachus, R.A., Trieu, W., Ro, D.K. and Keasling, J.D. (2007). Engineering Escherichia coli for production of functionalized terpenoids using plant P450s. *Nat. Chem. Biol.* 3, 274-277.

Croteau, R., Ketchum, R.E., Long, R.M., Kaspera, R. and Wildung, M.R. (2006). Taxol biosynthesis and molecular genetics. *Phytochem Rev* 5, 75-97.

Croteau, R.B., Davis, E.M., Ringer, K.L. and Wildung, M.R. (2005). (-)-Menthol biosynthesis and molecular genetics. *Naturwissenschaften* 92, 562-77.

Decorzant, R., Vial, C., Naf, F. and Whitesides, G.M. (1987). A Short Synthesis of Ambrox from Sclareol. *Tetrahedron* 43, 1871-1879.

Delaney, S.K., Orford, S.J., Martin-Harris, M. and Timmis, J.N. (2007). The fiber specificity of the cotton FSltp4 gene promoter is regulated by an AT-rich promoter region and the AT-hook transcription factor GhAT1. *Plant and Cell Physiology* 48, 1426-1437.

Diemer, F., Jullien, F., Faure, O., Moja, S., Colson, M., Matthys-Rochon, E. and Caissard, J.C. (1998). High efficiency transformation of peppermint (Mentha x piperita L.) with Agrobacterium tumefaciens. *Plant Sci.* 136, 101-108.

Ennajdaoui, H., Vachon, G., Giacalone, C., Besse, I., Sallaud, C., Herzog, M. and Tissier, A. (2010). Trichome specific expression of the tobacco (Nicotiana sylvestris) cembratrien-ol synthase genes is controlled by both activating and repressing cis-regions. *Plant Mol. Biol.* 73, 673-85.

Fei, Z., Tang, X., Alba, R.M., White, J.A., Ronning, C.M., Martin, G.B., Tanksley, S.D. and Giovannoni, J.J. (2004). Comprehensive EST analysis of tomato and comparative genomics of fruit ripening. *The Plant Journal* 40, 47-59.

Fridman, E., Wang, J., Iijima, Y., Froehlich, J.E., Gang, D.R., Ohlrogge, J. and Pichersky, E. (2005). Metabolic, genomic, and biochemical analyses of glandular trichomes from the wild tomato species Lycopersicon hirsutum identify a key enzyme in the biosynthesis of methylketones. *Plant Cell* 17, 1252-67.

Gang, D.R., Beuerle, T., Ullmann, P., Werck-Reichhart, D. and Pichersky, E. (2002a). Differential production of meta hydroxylated phenylpropanoids in sweet basil peltate glandular trichomes and leaves is controlled by the activities of specific acyltransferases and hydroxylases. *Plant Physiol.* 130, 1536-44.

Gang, D.R., Lavid, N., Zubieta, C., Chen, F., Beuerle, T., Lewinsohn, E., Noel, J.P. and Pichersky, E. (2002b). Characterization of phenylpropene O-methyltransferases from sweet basil: facile change of substrate specificity and convergent evolution within a plant O-methyltransferase family. *Plant Cell* 14, 505-19.

Gang, D.R., Wang, J., Dudareva, N., Nam, K.H., Simon, J.E., Lewinsohn, E. and Pichersky, E. (2001). An investigation of the storage and biosynthesis of phenylpropenes in sweet basil. *Plant Physiol.* 125, 539-55.

Gershenzon, J., Duffy, M.A., Karp, F. and Croteau, R. (1987). Mechanized techniques for the selective extraction of enzymes from plant epidermal glands. *Anal Biochem* 163, 159-64.

Gershenzon, J., Maffei, M. and Croteau, R. (1989). Biochemical and Histochemical Localization of Monoterpene Biosynthesis in the Glandular Trichomes of Spearmint (Mentha spicata). *Plant Physiol.* 89, 1351-7.

Gershenzon, J., McCaskill, D., Rajaonarivony, J.I., Mihaliak, C., Karp, F. and Croteau, R. (1992). Isolation of secretory cells from plant glandular trichomes and their use in biosynthetic studies of monoterpenes and other gland products. *Anal Biochem* 200, 130-8.

Gershenzon, J., McConkey, M.E. and Croteau, R.B. (2000). Regulation of monoterpene accumulation in leaves of peppermint. *Plant Physiol.* 122, 205-14.

Gilardon, E., Pocovi, M., Hernandez, C., Collavino, G. and Olsen, A. (2001). Role of 2-tridecanone and type VI glandular trichome on tomato resistance to Tuta absoluta. *Pesqui Agropecu Bras* 36, 929-933.

Graham, I.A., Besser, K., Blumer, S., Branigan, C.A., Czechowski, T., Elias, L., Guterman, I., Harvey, D., Isaac, P.G., Khan, A.M. et al. (2010). The Genetic Map of Artemisia annua L. Identifies Loci Affecting Yield of the Antimalarial Drug Artemisinin. *Science* 327, 328-331.

Guan, X.Y., Li, Q.J., Shan, C.M., Wang, S., Mao, Y.B., Wang, L.J. and Chen, X.Y. (2008). The HD-Zip IV gene GaHOX1 from cotton is a functional homologue of the Arabidopsis GLABRA2. *Physiol Plantarum* 134, 174-182.

Guitton, Y., Nicole, F., Moja, S., Valot, N., Legrand, S., Jullien, F. and Legendre, L. (2010). Differential accumulation of volatile terpene and terpene synthase mRNAs during lavender (Lavandula angustifolia and L. x intermedia) inflorescence development. *Physiol Plantarum* 138, 150-63.

Guo, Z., Severson, R.F. and Wagner, G.J. (1994). Biosynthesis of the diterpene cis-abienol in cell-free extracts of tobacco trichomes. *Arch. Biochem. Biophys.* 308, 103-8.

Gutierrez-Alcala, G., Calo, L., Gros, F., Caissard, J.C., Gotor, C. and Romero, L.C. (2005). A versatile promoter for the expression of proteins in glandular and non-glandular trichomes from a variety of plants. *Journal of Experimental Botany* 56, 2487-2494.

Harmer, S.H., Orford, S.O. and Timmis, J.T. (2002). Characterisation of six a-expansin genes in<SMALL> Gossypium hirsutum</SMALL> (upland cotton). *Mol. Genet. Genomics* 268, 1-9.

Hefner, J., Rubenstein, S.M., Ketchum, R.E., Gibson, D.M., Williams, R.M. and Croteau, R. (1996). Cytochrome P450-catalyzed hydroxylation of taxa-4(5),11(12)-diene to taxa-4(20),11(12)-dien-5alpha-ol: the first oxygenation step in taxol biosynthesis. *Chem Biol* 3, 479-89.

Heinz, K.M. and Zalom, F.G. (1995). Variation in Trichome-Based Resistance to Bemisia-Argentifolii (Homoptera, Aleyrodidae) Oviposition on Tomato. *Journal of Economic Entomology* 88, 1494-1502.

Ho, M.H., Saha, S., Jenkins, J.N. and Ma, D.P. (2010). Characterization and Promoter Analysis of a Cotton RING-Type Ubiquitin Ligase (E3) Gene. *Mol. Biotechnol.* 46, 140-148.

Hou, L., Liu, H., Li, J.B., Yang, X., Xiao, Y.H., Luo, M., Song, S.Q., Yang, G.W. and Pei, Y. (2008). SCFP, a novel fiber-specific promoter in cotton. *Chinese Sci Bull* 53, 2639-2645.

Hsu, C.Y., Creech, R.G., Jenkins, J.N. and Ma, D.P. (1999). Analysis of promoter activity of cotton lipid transfer protein gene LTP6 in transgenic tobacco plants. *Plant Sci.* 143, 63-70.

Iijima, Y., Davidovich-Rikanati, R., Fridman, E., Gang, D.R., Bar, E., Lewinsohn, E. and Pichersky, E. (2004a). The biochemical and molecular basis for the divergent patterns in the biosynthesis of terpenes and phenylpropenes in the peltate glands of three cultivars of basil. *Plant Physiol.* 136, 3724-36.

Iijima, Y., Gang, D.R., Fridman, E., Lewinsohn, E. and Pichersky, E. (2004b). Characterization of geraniol synthase from the peltate glands of sweet basil. *Plant Physiol.* 134, 370-9.

Jennewein, S., Long, R.M., Williams, R.M. and Croteau, R. (2004). Cytochrome p450 taxadiene 5alpha-hydroxylase, a mechanistically unusual monooxygenase catalyzing the first oxygenation step of taxol biosynthesis. *Chem Biol* 11, 379-87.

John, M.E. and Crow, L.J. (1992). GENE-EXPRESSION IN COTTON (GOSSYPIUM-HIRSUTUM L) FIBER - CLONING OF THE MESSENGER-RNAS. *Proc. Natl. Acad. Sci. USA* 89, 5769-5773.

John, M.E. and Keller, G. (1996). Metabolic pathway engineering in cotton: Biosynthesis of polyhydroxybutyrate in fiber cells. *Proceedings of the National Academy of Sciences of the United States of America* 93, 12768-12773.

Kandra, L. and Wagner, G.J. (1988). Studies of the site and mode of biosynthesis of tobacco trichome exudate components. *Arch. Biochem. Biophys.* 265, 425-32.

Keene, C.K. and Wagner, G.J. (1985). Direct demonstration of duvatrienediol biosynthesis in glandular heads of tobacco trichomes. *Plant Physiol.* 79, 1026-32.

Kennedy, G.G. (2003). Tomato, pests, parasitoids, and predators: tritrophic interactions involving the genus Lycopersicon. *Annu Rev Entomol* 48, 51-72.

Kim, H.J., Murai, N., Fang, D.D. and Triplett, B.A. (2011). Functional analysis of Gossypium hirsutum cellulose synthase catalytic subunit 4 promoter in transgenic Arabidopsis and cotton tissues. *Plant Sci.* 180, 323-332.

Koga, T., Aoki, Y., Hirose, T. and Nohira, H. (1998). Resolution of sclareolide as a key intermediate for the synthesis of Ambrox (R). *Tetrahedron-Asymmetr* 9, 3819-3823.

Kovacs, K., Zhang, L., Linforth, R.S.T., Whittaker, B., Hayes, C.J. and Fray, R.G. (2007). Redirection of carotenoid metabolism for the efficient production of taxadiene taxa-4(5),11(12)-diene in transgenic tomato fruit. *Transgenic Res.* 16, 121-126.

Kowalski, S.P., Eannetta, N.T., Hirzel, A.T. and Steffens, J.C. (1992). Purification and Characterization of Polyphenol Oxidase from Glandular Trichomes of Solanum berthaultii. *Plant Physiology* 100, 677-684.

Kuzma, L., Bruchajzer, E. and Wysokinska, H. (2008). Diterpenoid production in hairy root culture of Salvia sclarea L. *Zeitschrift Fur Naturforschung Section C-a Journal of Biosciences* 63, 621-624.

Kuzma, L., Skrzypek, Z. and Wysokinska, H. (2006). Diterpenoids and triterpenoids in hairy roots of Salvia sclarea. *Plant Cell Tiss Org* 84, 171-179.

Lane, A., Boecklemann, A., Woronuk, G.N., Sarker, L. and Mahmoud, S.S. (2010). A genomics resource for investigating regulation of essential oil production in Lavandula angustifolia. *Planta* 231, 835-845.

Lange, B.M. and Croteau, R. (1999). Genetic engineering of essential oil production in mint. *Current Opinion in Plant Biology* 2, 139-144.

Lattoo, S.K., Dhar, R.S., Dhar, A.K., Sharma, P.R. and Agarwal, S.G. (2006). Dynamics of essential oil biosynthesis in relation to inflorescence and glandular ontogeny in Salvia sclarea. *Flavour Frag J* 21, 817-821.

Lee, C.Y., Agrawal, D.C., Wang, C.S., Yu, S.M., Chen, J.J.W. and Tsay, H.S. (2008). T-DNA activation tagging as a tool to isolate Salvia miltiorrhiza transgenic lines for higher yields of tanshinones. *Planta Medica* 74, 780-786.

Lee, J., Burns, T., Light, G., Sun, Y., Fokar, M., Kasukabe, Y., Fujisawa, K., Maekawa, Y. and Allen, R. (2010). Xyloglucan endotransglycosylase/hydrolase genes in cotton and their role in fiber elongation. *Planta* 232, 1191-1205.

Lee, J.J., Woodward, A.W. and Chen, Z.J. (2007). Gene expression changes and early events in cotton fibre development. *Ann Bot* 100, 1391-401.

Legrand, S., Valot, N., Nicole, F., Moja, S., Baudino, S., Jullien, F., Magnard, J.L., Caissard, J.C. and Legendre, L. (2010). One-step identification of conserved miRNAs, their targets, potential transcription factors and effector genes of complete secondary metabolism pathways after 454 pyrosequencing of calyx cDNAs from the Labiate Salvia sclarea L. *Gene* 450, 55-62.

Li, C.H., Zhu, Y.Q., Meng, Y.L., Wang, J.W., Xu, K.X., Zhang, T.Z. and Chen, X.Y. (2002a). Isolation of genes preferentially expressed in cotton fibers by cDNA filter arrays and RT-PCR. *Plant Sci.* 163, 1113-1120.

Li, L., Wang, X.L., Huang, G.Q. and Li, X.B. (2007). Molecular characterization of cotton GhTUA9 gene specifically expressed in fibre and involved in cell elongation. *J. Exp. Bot.* 58, 3227-3238.

Li, X.B., Cai, L., Cheng, N.H. and Liu, J.W. (2002b). Molecular characterization of the cotton GhTUB1 gene that is preferentially expressed in fiber. *Plant Physiol.* 130, 666-674.

Li, X.B., Fan, X.P., Wang, X.L., Cai, L. and Yang, W.C. (2005). The cotton ACTIN1 gene is functionally expressed in fibers and participates in fiber elongation. *Plant Cell* 17, 859-875.

Liu, H.C., Creech, R.G., Jenkins, J.N. and Ma, D.P. (2000). Cloning and promoter analysis of the cotton lipid transfer protein gene Ltp3(1). *Biochim Biophys Acta* 1487, 106-11.

Liu, S.Q., Tian, N., Li, J., Huang, J.N. and Liu, Z.H. (2009). Isolation and Identification of Novel Genes Involved in Artemisinin Production from Flowers of Artemisia annua Using Suppression Subtractive Hybridization and Metabolite Analysis. *Planta Medica* 75, 1542-1547.

Luckwill, L.C. (1943). The Genus Lycopersicon: A Historical, Biological, and Taxonomic Survey of the Wild and Cultivated Tomatoes: Aberdeen University Press, Aberdeen, Scotland.

Luo, M., Xiao, Y.H., Li, X.B., Lu, X.F., Deng, W., Li, D., Hou, L., Hu, M.Y., Li, Y. and Pei, Y. (2007). GhDET2, a steroid 5 alpha-reductase, plays an important role in cotton fiber cell initiation and elongation. *Plant J.* 51, 419-430.

Lupien, S., Karp, F., Wildung, M. and Croteau, R. (1999). Regiospecific cytochrome P450 limonene hydroxylases from mint (Mentha) species: cDNA isolation, characterization, and functional expression of (-)-4S-limonene-3-hydroxylase and (-)-4S-limonene-6-hydroxylase. *Arch. Biochem. Biophys.* 368, 181-92.

Ma, D.M., Pu, G.B., Lei, C.Y., Ma, L.Q., Wang, H.H., Guo, Y.W., Chen, J.L., Du, Z.G., Wang, H., Li, G.F. et al. (2009). Isolation and Characterization of AaWRKY1, an Artemisia annua Transcription Factor that Regulates the Amorpha-4,11-diene Synthase Gene, a Key Gene of Artemisinin Biosynthesis. *Plant and Cell Physiology* 50, 2146-2161.

Ma, D.P., Tan, H., Si, Y., Creech, R.G. and Jenkins, J.N. (1995). DIFFERENTIAL EXPRESSION OF A LIPID TRANSFER PROTEIN GENE IN COTTON FIBER. *Biochim. Biophys. Acta-Lipids Lipid Metab.* 1257, 81-84.

Machado, A., Wu, Y., Yang, Y., Llewellyn, D.J. and Dennis, E.S. (2009). The MYB transcription factor GhMYB25 regulates early fibre and trichome development. *Plant J.* 59, 52-62.

Maluf, W.R., Silva, V.D., Cardoso, M.D., Gomes, L.A.A., Neto, A.C.G., Maciel, G.M. and Nizio, D.A.C. (2010). Resistance to the South American tomato pinworm Tuta absoluta in high acylsugar and/or high zingiberene tomato genotypes. *Euphytica* 176, 113-123.

Marillonnet, S., Thoeringer, C., Kandzia, R., Klimyuk, V. and Gleba, Y. (2005). Systemic Agrobacterium tumefaciens-mediated transfection of viral replicons for efficient transient expression in plants. *Nat Biotechnol* 23, 718-23.

Martin, V.J.J., Pitera, D.J., Withers, S.T., Newman, J.D. and Keasling, J.D. (2003). Engineering a mevalonate pathway in Escherichia coli for production of terpenoids. *Nat. Biotechnol.* 21, 796-802.

Martres, P., Perfetti, P., Zahra, J.P., Waegell, B., Giraudi, E. and Petrzilka, M. (1993). A Short and Efficient Synthesis of (-)-Ambrox(R) from (-)-Sclareol Using a Ruthenium Oxide Catalyzed Key Step. *Tetrahedron Lett.* 34, 629-632.

McConkey, M.E., Gershenzon, J. and Croteau, R.B. (2000). Developmental regulation of monoterpene biosynthesis in the glandular trichomes of peppermint. *Plant Physiol.* 122, 215-24.

McDowell, E.T., Kapteyn, J., Schmidt, A., Li, C., Kang, J.-H., Descour, A., Shi, F., Larson, M., Schilmiller, A., An, L. et al. (2011). Comparative Functional Genomic Analysis of Solanum Glandular Trichome Types. *Plant Physiol.* 155, 524-539.

Mercke, P., Bengtsson, M., Bouwmeester, H.J., Posthumus, M.A. and Brodelius, P.E. (2000). Molecular cloning, expression, and characterization of amorpha-4,11-diene synthase, a key enzyme of artemisinin biosynthesis in Artemisia annua L. *Archives of Biochemistry and Biophysics* 381, 173-180.

Michailidis, G., Argiriou, A., Darzentas, N. and Tsaftaris, A. (2009). Analysis of xyloglucan endotransglycosylase/hydrolase (XTH) genes from allotetraploid (Gossypium hirsutum) cotton and its diploid progenitors expressed during fiber elongation. *Journal of Plant Physiology* 166, 403-416.

Mishiba, K.I., Ishikawa, K., Tsujii, O. and Mii, M. (2000). Efficient transformation of lavender (Lavandula latifolia Medicus) mediated by Agrobacterium. *J Hortic Sci Biotech* 75, 287-292.

Moulines, J., Bats, J.P., Lamidey, A.M. and Da Silva, N. (2004). About a practical synthesis of Ambrox (R) from sclareol: a new preparation of a ketone key intermediate and a close look at its Baeyer-Villiger oxidation. *Helv Chim Acta* 87, 2695-2705.

Moulines, J., Lamidey, A.M. and Desvergnes-Breuil, V. (2001). A practical synthesis of Ambrox (R) from sclareol using no metallic oxidant. *Synth. Commun.* 31, 749-758.

Munoz-Bertomeu, J., Ros, R., Arrillaga, I. and Segura, J. (2008). Expression of spearmint limonene synthase in transgenic spike lavender results in an altered monoterpene composition in developing leaves. *Metabolic Engineering* 10, 166-177.

Munoz-Bertomeu, J., Sales, E., Ros, R., Arrillaga, I. and Segura, J. (2007). Up-regulation of an N-terminal truncated 3-hydroxy-3-methylglutaryl CoA reductase enhances production of essential oils and sterols in transgenic Lavandula latifolia. *Plant Biotechnol. J.* 5, 746-758.

Nebauer, S.G., Arrillaga, I., del Castillo-Agudo, L. and Segura, J. (2000). Agrobacterium tumefaciens-mediated transformation of the aromatic shrub Lavandula latifolia. *Molecular Breeding* 6, 539-552.

Newman, J.D., Marshall, J., Chang, M., Nowroozi, F., Paradise, E., Pitera, D., Newman, K.L. and Keasling, J.D. (2006). High-level production of amorpha-4,11-diene in a two-phase partitioning bioreactor of metabolically engineered Escherichia coli. *Biotechnology and Bioengineering* 95, 684-691.

Olofsson, L., Engstrom, A., Lundgren, A. and Brodelius, P.E. (2011). Relative expression of genes of terpene metabolism in different tissues of Artemisia annua L. *BMC Plant Biol.* 11.

Olsson, M.E., Olofsson, L.M., Lindahl, A.L., Lundgren, A., Brodelius, M. and Brodelius, P.E. (2009). Localization of enzymes of artemisinin biosynthesis to the apical cells of glandular secretory trichomes of Artemisia annua L. *Phytochemistry* 70, 1123-1128.

Orford, S.J., Carney, T.J., Olesnicky, N.S. and Timmis, J.N. (1999). Characterisation of a cotton gene expressed late in fibre cell elongation. *Theor. Appl. Genet.* 98, 757-764.

Orford, S.J. and Timmis, J.M. (1997). Abundant mRNAs specific to the developing cotton fibre. *Theor. Appl. Genet.* 94, 909-918.

Orford, S.J. and Timmis, J.N. (1995). SPECIFIC GENES EXPRESSED DURING COTTON FIBER DEVELOPMENT. *Journal of Cellular Biochemistry*, 447-447.

Orford, S.J. and Timmis, J.N. (1998). Specific expression of an expansin gene during elongation of cotton fibres. *Biochim. Biophys. Acta-Gene Struct. Expression* 1398, 342-346.

Payne, W. (1978). A glossary of plant hair terminology. *Brittonia* 30, 239-255.

Peters, R.J. (2010). Two rings in them all: the labdane-related diterpenoids. *Nat. Prod. Rep.* 27, 1521-30.

Poirier, Y., Dennis, D.E., Klomparens, K. and Somerville, C. (1992). Polyhydroxybutyrate, a Biodegradable Thermoplastic, Produced in Transgenic Plants. *Science* 256, 520-523.

Pu, L., Li, Q., Fan, X.P., Yang, W.C. and Xue, Y.B. (2008). The R2R3 MYB Transcription Factor GhMYB109 Is Required for Cotton Fiber Development. *Genetics* 180, 811-820.

Rajaonarivony, J.I., Gershenzon, J. and Croteau, R. (1992). Characterization and mechanism of (4S)-limonene synthase, a monoterpene cyclase from the glandular trichomes of peppermint (Mentha x piperita). *Arch. Biochem. Biophys.* 296, 49-57.

Rasoulpour, R. and Izadpanah, K. (2007). Characterisation of cineraria strain of Tomato yellow ring virus from Iran. *Australasian Plant Pathology* 36, 286-294.

Rinehart, J.A., Petersen, M.W. and John, M.E. (1996). Tissue-specific and developmental regulation of cotton gene FbL2A - Demonstration of promoter activity in transgenic plants. *Plant Physiol.* 112, 1331-1341.

Ro, D.K., Paradise, E.M., Ouellet, M., Fisher, K.J., Newman, K.L., Ndungu, J.M., Ho, K.A., Eachus, R.A., Ham, T.S., Kirby, J. et al. (2006). Production of the antimalarial drug precursor artemisinic acid in engineered yeast. *Nature* 440, 940-943.

Rontein, D., Onillon, S., Herbette, G., Lesot, A., Werck-Reichhart, D., Sallaud, C. and Tissier, A. (2008). CYP725A4 from yew catalyzes complex structural rearrangement of taxa-4(5),11(12)-diene into the cyclic ether 5(12)-oxa-3(11)-cyclotaxane. *J. Biol. Chem.* 283, 6067-75.

Ruan, M.B., Liao, W.B., Zhang, X.C., Yu, X.L. and Peng, M. (2009). Analysis of the cotton sucrose synthase 3 (Sus3) promoter and first intron in transgenic Arabidopsis. *Plant Sci.* 176, 342-351.

Rubinstein, G. and Czosnek, H. (1997). Long-term association of tomato yellow leaf curl virus with its whitefly vector Bemisia tabaci: effect on the insect transmission capacity, longevity and fecundity. *J. Gen. Virol.* 78 (Pt 10), 2683-9.

Sallaud, C., Rontein, D., Onillon, S., Jabes, F., Duffe, P., Giacalone, C., Thoraval, S., Escoffier, C., Herbette, G., Leonhardt, N. et al. (2009). A novel pathway for sesquiterpene biosynthesis from Z,Z-farnesyl pyrophosphate in the wild tomato Solanum habrochaites. *Plant Cell* 21, 301-17.

Schilmiller, A.L., Last, R.L. and Pichersky, E. (2008). Harnessing plant trichome biochemistry for the production of useful compounds. *Plant J.* 54, 702-11.

Schilmiller, A.L., Schauvinhold, I., Larson, M., Xu, R., Charbonneau, A.L., Schmidt, A., Wilkerson, C., Last, R.L. and Pichersky, E. (2009). Monoterpenes in the glandular trichomes of tomato are synthesized from a neryl diphosphate precursor rather than geranyl diphosphate. *Proc. Natl. Acad. Sci. USA* 106, 10865-70.

Schmiderer, C., Grassi, P., Novak, J., Weber, M. and Franz, C. (2008). Diversity of essential oil glands of clary sage (Salvia sclarea L., Lamiaceae). *Plant Biology* 10, 433-440.

Serna, L. and Martin, C. (2006). Trichomes: different regulatory networks lead to convergent structures. *Trends Plant Sci* 11, 274-80.

Shangguan, X.X., Xu, B., Yu, Z.X., Wang, L.J. and Chen, X.Y. (2008). Promoter of a cotton fibre MYB gene functional in trichomes of Arabidopsis and glandular trichomes of tobacco. *Journal of Experimental Botany* 59, 3533-3542.

Shepherd, R.W., Bass, W.T., Houtz, R.L. and Wagner, G.J. (2005). Phylloplanins of tobacco are defensive proteins deployed on aerial surfaces by short glandular trichomes. *Plant Cell* 17, 1851-1861.

Slocombe, S.P., Schauvinhold, I., McQuinn, R.P., Besser, K., Welsby, N.A., Harper, A., Aziz, N., Li, Y., Larson, T.R., Giovannoni, J. et al. (2008). Transcriptomic and reverse genetic analyses of branched-chain fatty acid and acyl sugar production in Solanum pennellii and Nicotiana benthamiana. *Plant Physiol.* 148, 1830-46.

Snyder, J.C., Simmons, A.M. and Thacker, R.R. (1998). Attractancy and ovipositional response of adult Bemisia argentifolii (Homoptera : Aleyrodidae) to type IV trichome density on leaves of Lycopersicon hirsutum grown in three day-length regimes. *J Entomol Sci* 33, 270-281.

Suo, J., Liang, X., Pu, L., Zhang, Y. and Xue, Y. (2003). Identification of GhMYB109 encoding a R2R3 MYB transcription factor that expressed specifically in fiber initials and elongating fibers of cotton (Gossypium hirsutum L.). *Biochim Biophys Acta* 1630, 25-34.

Taliercio, E. (2011). Characterization of an ADP-glucose pyrophosphorylase small subunit gene expressed in developing cotton (<i>Gossypium hirsutum</i>) fibers. *Molecular Biology Reports* 38, 2967-2973.

Teoh, K.H., Polichuk, D.R., Reed, D.W. and Covello, P.S. (2009). Molecular cloning of an aldehyde dehydrogenase implicated in artemisinin biosynthesis in Artemisia annua. *Botany* 87, 635-642.

Thipyapong, P., Joel, D.M. and Steffens, J.C. (1997). Differential Expression and Turnover of the Tomato Polyphenol Oxidase Gene Family during Vegetative and Reproductive Development. *Plant Physiol.* 113, 707-718.

Turner, G., Gershenzon, J., Nielson, E.E., Froehlich, J.E. and Croteau, R. (1999). Limonene synthase, the enzyme responsible for monoterpene biosynthesis in peppermint, is localized to leucoplasts of oil gland secretory cells. *Plant Physiol.* 120, 879-886.

Udall, J.A., Swanson, J.M., Haller, K., Rapp, R.A., Sparks, M.E., Hatfield, J., Yu, Y.S., Wu, Y.R., Dowd, C., Arpat, A.B. et al. (2006). A global assembly of cotton ESTs. *Genome Res.* 16, 441-450.

Uhrig, J.F. and Hulskamp, M. (2010). Trichome development in Arabidopsis. *Methods Mol Biol* 655, 77-88.

van Der Hoeven, R.S., Monforte, A.J., Breeden, D., Tanksley, S.D. and Steffens, J.C. (2000). Genetic control and evolution of sesquiterpene biosynthesis in Lycopersicon esculentum and L. hirsutum. *Plant Cell* 12, 2283-94.

Vendramim, J.D., de Souza, A.P. and Ongarelli, M.D. (2009). Oviposition Behavior of the Silverleaf Whitefly Bemisia tabaci Biotype B on Tomato. *Neotrop. Entomol.* 38, 126-132.

Wagner, G.J., Wang, E. and Shepherd, R.W. (2004). New approaches for studying and exploiting an old protuberance, the plant trichome. *Ann Bot* 93, 3-11.

Wallaart, T.E., Bouwmeester, H.J., Hille, J., Poppinga, L. and Maijers, N.C.A. (2001). Amorpha-4,11-diene synthase: cloning and functional expression of a key enzyme in the biosynthetic pathway of the novel antimalarial drug artemisinin. *Planta* 212, 460-465.

Wang, E. and Wagner, G.J. (2003). Elucidation of the functions of genes central to diterpene metabolism in tobacco trichomes using posttranscriptional gene silencing. *Planta* 216, 686-91.

Wang, E., Wang, R., DeParasis, J., Loughrin, J.H., Gan, S. and Wagner, G.J. (2001). Suppression of a P450 hydroxylase gene in plant trichome glands enhances natural-product-based aphid resistance. *Nat. Biotechnol.* 19, 371-4.

Wang, E.M., Gan, S.S. and Wagner, G.J. (2002). Isolation and characterization of the CYP71D16 trichome-specific promoter from Nicotiana tabacum L. *J. Exp. Bot.* 53, 1891-1897.

Wang, S., Wang, J.W., Yu, N., Li, C.H., Luo, B., Gou, J.Y., Wang, L.J. and Chen, X.Y. (2004). Control of plant trichome development by a cotton fiber MYB gene. *Plant Cell* 16, 2323-34.

Wang, W., Wang, Y.J., Zhang, Q., Qi, Y. and Guo, D.J. (2009). Global characterization of Artemisia annua glandular trichome transcriptome using 454 pyrosequencing. *BMC Genomics* 10, -.

Weathers, P.J., Arsenault, P.R., Covello, P.S., McMickle, A., Teoh, K.H. and Reed, D.W. (2011). Artemisinin production in Artemisia annua: studies in planta and results of a novel delivery method for treating malaria and other neglected diseases. *Phytochem. Rev.* 10, 173-183.

Weller, S.C., Niu, X., Lin, K., Hasegawa, P.M. and Bressan, R.A. (1998). Transgenic peppermint (Mentha x piperita L.) plants obtained by cocultivation with Agrobacterium tumefaciens. *Plant Cell Rep.* 17, 165-171.

Wu, A.-M., Hu, J. and Liu, J.-Y. (2009). Functional analysis of a cotton cellulose synthase A4 gene promoter in transgenic tobacco plants. *Plant Cell Rep.* 28, 1539-1548.

Wu, A.M., Lv, S.Y. and Liu, J.Y. (2007). Functional analysis of a cotton glucuronosyltransferase promoter in transgenic tobaccos. *Cell Research* 17, 174-183.

Wu, S., Schalk, M., Clark, A., Miles, R.B., Coates, R. and Chappell, J. (2006). Redirection of cytosolic or plastidic isoprenoid precursors elevates terpene production in plants. *Nat Biotechnol* 24, 1441-7.

Wu Y, X.W., Huang G, Gong S, Li J, Qin Y, Li X. (2009). Expression and localization of GhH6L, a putative classical arabinogalactan protein in cotton (Gossypium hirsutum). *Acta Biochim Biophys Sin* 41, 495-503.

Wust, M., Little, D.B., Schalk, M. and Croteau, R. (2001). Hydroxylation of limonene enantiomers and analogs by recombinant (-)-limonene 3- and 6-hydroxylases from mint (Mentha) species: evidence for catalysis within sterically constrained active sites. *Arch. Biochem. Biophys.* 387, 125-36.

Xu, X., Wu, M., Zhao, Q., Li, R., Chen, J., Ao, G. and Yu, J. (2007). Designing and transgenic expression of melanin gene in tobacco trichome and cotton fiber. *Plant Biol (Stuttg)* 9, 41-8.

Yan, Y.P. and Wang, Z.Z. (2007). Genetic transformation of the medicinal plant Salvia miltiorrhiza by Agrobacterium tumefaciens-mediated method. *Plant Cell Tiss Org* 88, 175-184.

Yang, S.S., Cheung, F., Lee, J.J., Ha, M., Wei, N.E., Sze, S.H., Stelly, D.M., Thaxton, P., Triplett, B., Town, C.D. et al. (2006). Accumulation of genome-specific transcripts, transcription factors and phytohormonal regulators during early stages of fiber cell development in allotetraploid cotton. *Plant J.* 47, 761-775.

Yu, G., Nguyen, T.T.H., Guo, Y., Schauvinhold, I., Auldridge, M.E., Bhuiyan, N., Ben-Israel, I., Iijima, Y., Fridman, E., Noel, J.P. et al. (2010). Enzymatic Functions of Wild Tomato Methylketone Synthases 1 and 2. *Plant Physiol.* 154, 67-77.

Yu, H., Kowalski, S.P. and Steffens, J.C. (1992). Comparison of polyphenol oxidase expression in glandular trichomes of solanum and lycopersicon species. *Plant Physiol.* 100, 1885-90.

Zhang, D., Hrmova, M., Wan, C.-H., Wu, C., Balzen, J., Cai, W., Wang, J., Densmore, L.D., Fincher, G.B., Zhang, H. et al. (2004). Members of a New Group of Chitinase-Like Genes are Expressed Preferentially in Cotton Cells with Secondary Walls. *Plant Mol. Biol.* 54, 353-372.

Zhang, F., Zuo, K.J., Zhang, J.Q., Liu, X.A., Zhang, L.D., Sun, X.F. and Tang, K.X. (2010). An L1 box binding protein, GbML1, interacts with GbMYB25 to control cotton fibre development. *J. Exp. Bot.* 61, 3599-3613.

Zhang, Y., Teoh, K.H., Reed, D.W., Maes, L., Goossens, A., Olson, D.J.H., Ross, A.R.S. and Covello, P.S. (2008). The molecular cloning of artemisinic aldehyde Delta 11(13) reductase and its role in glandular trichome-dependent biosynthesis of artemisinin in Artemisia annua. *J. Biol. Chem.* 283, 21501-21508.

Zhang, Y.S., Nowak, G., Reed, D.W. and Covello, P.S. (2011). The production of artemisinin precursors in tobacco. *Plant Biotechnol. J.* 9, 445-454.

Zhang, Y.S., Teoh, K.H., Reed, D.W. and Covello, P.S. (2009). Molecular cloning and characterization of Dbr1, a 2-alkenal reductase from Artemisia annua. *Botany* 87, 643-649.

6

Comparative Metabolomics of Transgenic Tobacco Plants (*Nicotiana tabacum* var. *Xanthi*) Reveals Differential Effects of Engineered Complete and Incomplete Flavonoid Pathways on the Metabolome

Corey D. Broeckling[2], Ke-Gang Li[1] and De-Yu Xie[1]
[1]North Carolina State University, Department of Plant Biology, Raleigh, NC
[2]Proteomics and Metabolomics Facility, Colorado State University, Fort Collins, CO
USA

1. Introduction

Anthocyanins and proanthocyanidins (PAs) are two groups of end products of the plant flavonoid pathway (Fig. 1). Biochemical and genetic evidences have demonstrated that they share the same upstream pathway beginning with phenylalanine through a series of enzymatic reaction to anthocyanidins. Anthocyanidins are either modified by glycosylation, methylation, or other reactions to form diverse anthocyanins (Springob et al. 2003) or catalyzed into flavan-3-ols by an anthocyanidin reductase (ANR) (Xie et al. 2003; Xie et al. 2006). In addition, leucoanthocyanidin reductase (LAR) has been enzymatically demonstrated to catalyze leucoanthocyanidins into catechin (Tanner et al. 2003). To date, whether or not this branch catalyzed by LAR exists in plants still remains open to genetic studies.

Production of Anthocyanin Pigment 1 (*PAP1*) encodes a R2R3-MYB transcription factor and is a master regulator of anthocyanin biosynthesis in *Arabidopsis thaliana* (Fig. 1) (Borevitz et al. 2000; Tohge et al. 2005; Xie et al. 2006). The constitutive expression of *PAP1* in *Arabidopsis* and tobacco resulted in massive accumulation of targeted anthocyanins and numerous other phenylpropanoid compounds (Borevitz et al. 2000; Tohge et al. 2005). Our recent experiments revealed that the regulatory function of *PAP1* on anthocyanin biosynthesis is closely associated with cellular specificity in tissues. In transgenic tobacco leaves of *PAP1* and *pap1-D Arabidopsis* leaves, anthocyanins accumulated in epidermal cells and parenchymal cells of vascular bundles in veins (Shi and Xie 2010; Xie et al. 2006). It was interesting that the overexpression of *PAP1* led to massive accumulation of anthocyanins in transgenic tobacco leaf trichomes (Xie et al. 2006) but not in leaf trichomes of *pap1-D Arabidopsis thaliana* (Shi and Xie 2010). *In vitro* separation of red cells from other cells has demonstrated that the regulatory function of *PAP1* expression can be inherited by cell culture developed from mother plants (Zhou et al. 2008). Both no anthocyanin-producing white cells and anthocyanin-producing red cells were obtained from transgenic *PAP1* tobacco leaves. Transcript analysis showed that the level of the *PAP1* overexpression in these two types of cells were similar, indicating that the regulatory function of the *PAP1* expression was dependent upon cell types. (Zhou et al. 2008). The regulatory function of *PAP1* is also highly controlled by environmental factors. Although the *pap1-D Arabidopsis* plants over express *PAP1* leading to high anthocyanin pigmentation in most of tissues (Borevitz et al. 2000), when growth conditions were changed, anthocyanin levels and composition were dramatically altered in leaves of *pap1-D* plants (Rowan et al. 2009; Shi and Xie 2010; Tohge et al. 2005). In the same nutrition condition, high light conditions increased anthocyanin levels in leaves of *pap1-D* plants; under the same light condition, high nitrogen nutrition conditions increased anthocyanin levels in leaves of *pap1-D* plants; although in these conditions, the expression levels of *PAP1* were similar (Shi and Xie 2010). These inconsistent relationships between the expression levels of *PAP1* and anthocyanin levels were likely associated with other transcription factors involved in anthocyanin biosynthesis. PAP1 (MYB75), TT8 (transparent testa 8) (bHLH) / GL3 (glabra 3) (bHLH) and TTG1 (transparent testa glabra 1) (WD40) have been demonstrated to form a regulatory complex of MYB-bHLH-WD40 (MBW) controlling anthocyanin biosynthesis in *Arabidopsis* (Gonzalez et al. 2009; Ramsay and Glover 2005). We recently determined that the PAP1-TT8/GL3-TTG1 complex independently regulated anthocyanin biosynthesis in *pap1-D* cells (Shi and Xie 2011). In addition, there are other MBW regulatory complexes controlling anthocyanin biosynthesis in *Arabidopsis* (Gonzalez et al. 2009). In different growth conditions, the regulatory function of *PAP1* is likely essentially dependent upon these regulatory complexes.

ANR is a NADPH/NADH-dependent flavonoid reductase converting anthocyanidins to flavan-3-ols (e.g. epicatechin) and PAs (Fig. 1) (Xie et al. 2003). ANR is encoded by a *BANYULS* gene that was first cloned from young seeds of *Arabidopsis* (Devic et al. 1999). Its homologs were cloned from different species including a model legume plant *Medicago truncatula* (Xie et al. 2003). The constitutively ectopic expression of *ANR* in tobacco showed the loss of anthocyanins in flowers, in which ANR competitively catalyzed anthocyanidins into flavan-3-ols (e.g. epicatechin) and PAs (Xie et al. 2003). The transgenic vegetative tissues including leaf and stem tissues also expressed *ANR*, but failed to form epicatechin and PAs due to the absence of anthocyanidins (Xie et al. 2003; Xie et al. 2006). To establish a complete pathway of PAs in ANR transgenic leaves and stems, *PAP1* were ectopically expressed in them. The co-expression of *ANR* and *PAP1* produced flavan-3-ols and PAs in leaves and stems of transgenic tobacco plants (Xie et al. 2006). This result demonstrated that the expression of *ANR* alone resulted in production of ANR protein in transgenic leaves and stems and these tissues contained an incomplete PA pathway. This result also demonstrated that the overexpression of *PAP1* provided substrates for ANR; thus formed an effective platform for metabolic engineering of flavan-3-ols and PAs.

Whether the ectopic expression of *PAP1*, *ANR*, and *PAP1::ANR*, which form different pathways in transgenic tobacco vegetative tissues, can impact other metabolisms beyond the flavonoid pathway remains unknown. In this study, metabolic profiles of wild-type (WT) and transgenic tobacco plants expressing *PAP1*, *ANR*, *and PAP1::ANR* were examined by GC-MS analysis, showing that these three transgenic events differentially altered accumulation patterns of both targeted and non-targeted metabolites beyond anthocyanins and PAs in both transgenic leaves and stems

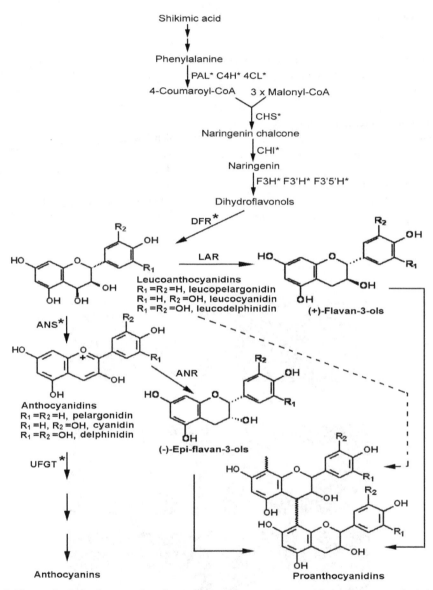

Fig. 1. Biosynthetic pathways of anthocyanins and proanthocyanidins starting with shikimic
acid. Enzymes include: PAL, L-phenylalanine ammonia-lyase; C4H, cinnamate-4-
hydroxylase; 4CL, 4-coumarate: CoA-ligase; CHS, chalcone synthase; CHI, chalcone
isomerase; F3H, flavanone-3-hydroxylase; F3'H, flavonoid 3'-hydroxylase; F3'5'H, flavonoid
3',5'-hydroxylase; DFR, dihydroflavonol reductase; ANS, anthocyanidin synthase; ANR,
anthocyanidin reductase; UFGT: uridine diphosphate glucose-flavonoid 3-O-
glucosyltransferase. Two arrows between shikimic acid and phenylalanine means there are
multiple steps. Asterisks (*) indicate steps known to be up-regulated by *PAP1* expression.

2. Materials and methods

2.1 Plant growth

Plant materials used for this experiment included *PAP1, ANR* and *PAP1:: ANR* (F1 progeny) transgenic plants as well as wild-type (WT) plants. WT plants was the control line #3, which has been used as control plants to characterize *PAP1* gene function in metabolic engineering of PAs (Xie et al. 2006). *PAP1* transgenic plants containing an engineered pathway of anthocyanins were a homozygotic one from the line #292 (Xie et al. 2006). The *ANR* transgenic plants were derived from the line # B-21, which has only one copy of the *MtANR* (*Medicago truncatula* ANR) transgene and produces PAs in flowers, (Xie et al. 2003), but contains an incomplete pathway of PAs in leaves and stems. The *PAP1::ANR* transgenic plants were derived from the line # P-B-13 (PAP1 x ANR) that was obtained by crossing the line # B-21 with the line #292 of *PAP1* transgenic plants. This line produced a relatively high level of PAs in leaves and stems resulting from an engineered PA pathway (Xie et al. 2006).

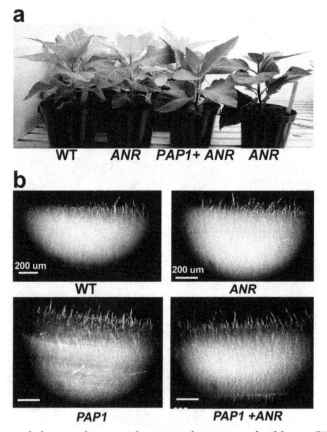

Fig. 2. Phenotypes of plants and stem trichomes. a phenotypes of wild-type (WT) and transgenic tobacco plants; b trichomes from stems; WT: wild type; *ANR, PAP1,* and *PAP1 + ANR: ANR, PAP1,* and *PAP1::ANR* transgenic plants.

Ten individual plants from each of four genetic background lines were vegetatively propagated by cutting. Finally, 37 plants (10 for wild-type and *PAP1* and *PAP1::ANR* transgenic plants, respectively, and 7 for *ANR* transgenic plants) were grown in a growth chamber for metabolic profiling. To minimize effects of physical factors (e.g. temperature, lighting, and humidity) on *in vivo* biochemical variability, plants were grown in pots with a size of three inches in radius and eight inches in height in a growth chamber at a constant temperature of 25°C, 70% humidity, 16:8 hr light: dark cycle, and under fluorescent epi-illumination. The possible effects of shading on plant growth were reduced by rotating the pots. The growth of all plants was morphologically similar (Fig. 2 a).

Sixty-two day old plants, which were approximately 30 cm in height with 10-13 leaves, were used for metabolome analysis. Plant tissues were harvested directly into liquid nitrogen. Plant tissues collected included old leaves, young leaves, and stems. Old leaves were defined as those that were fully expanded, 15-20 cm in length, from each individual plant, while young leaves were defined as those with 5-10 cm in length, not yet fully expanded. Old leaves, young leaves or stems from a single plant was pooled to create one sample, thus 111 samples in total were obtained from 37 plants. The frozen samples were stored at -80°C, freeze-dried at -80°C, and then ground into a fine powder, which was stored at -80°C until use.

2.2 Tissue extraction and GC-MS analyses

A polar and a lipophilic extraction for each sample were profiled in this experiment. Metabolites were extracted from lyophilized plant material as described previously (Broeckling et al. 2005). Briefly, freeze-dried homogenized tissue (6.0 +/- 0.05 mg) was weighed into a vial (4.0 ml) for extraction of metabolites by adding $CHCl_3$ (1.5 ml) that contained an internal standard (heptadecanoic acid methyl ester). Vials were thoroughly vortexed and incubated at 50°C for 45 min, followed by addition of HPLC grade H_2O (1.5 ml) containing a second internal standard (ribitol). The biphasic system was thoroughly vortexed and then incubated for an additional 45 min at 50°C. The sample was then centrifuged at 3000 x g for 30 min at 4°C. One ml of each phase was collected and the solvent was then evaporated either using speed vacuum (aqueous phase) or dried under liquid nitrogen stream ($CHCl_3$ phase). The polar phase extraction was methoximated in 120 μl of 15 μg/ μl methoxyamine HCl in pyridine for 120 min at 50°C and then trimethylsilyalted by adding 120 μl of MSTFA + 1% TMCS (Pierce Biotechnology, Rockford, IL, USA) and incubating for 60 min at 50°C. The non-polar phase extraction was derivitized in 100 μl of 50% MSTFA + 1% TMCS in pyridine at 50°C for 60 min. One μl of each was injected onto an Agilent 6890 GC coupled to a 5973 MS. The polar sample was split at 15:1 and the non-polar sample split 1:1. The oven program was 80°C (2 min) and ramped at 5°C/min to 315°C (12 min). Separation was performed on a 60m DB-5MS (J&W Scientific – 0.25 mm ID and 0.25 μM film thickness) at a flow rate of 1.0 ml/min. Metabolites were identified by comparison to a library of electron impact mass spectra and GC retention time as described previously (Broeckling et al. 2005). Identification was performed by using AMDIS deconvolution and identification software (NIST). The same GC separation parameters were used for obtaining positive and negative-ion chemical ionization spectra for tentative identification of the cembratienols using methane as the ionization gas on an Agilent 5973N MS. Quantification was performed as described previously (Broeckling et al. 2005).

2.3 Data processing, statistical analysis and heatmap

Peak detection and deconvolution were performed with AMDIS (Halket et al. 1999) for 2-3 samples of each treatment. Resultant peak lists were imported and compiled in MET-IDEA (Broeckling et al. 2006) and then used to extract quantitative peak area values for polar and non-polar metabolites. Redundant peaks were removed from the dataset, peak area values were scaled to mean zero and standard deviation 1.0. The resulting data matrix was statistically analyzed with discriminant function analysis, ANOVA, and principal component analysis in JMP (SAS institute, Cary, North Carolina). One-Way ANOVA was performed to extract significant different levels of metabolites (p-values < 0.05), then followed by Tukey's HSD pos-hoc analysis to compare all pair-wise mean difference (p-value < 0.05). To visualize metabolite accumulation patterns and the effect of transgenes on metabolite levels in tissues, heatmaps were established from excel using vision functions in macro.

2.4 Histological analysis of anthocyanin accumulation

Both young leaves and stems were dissected to 10-20 μm in thickness by hands and cellular localization of anthocyanin accumulation was examined with light microscope as described previously (Xie et al. 2006).

3. Results

3.1 Plant growth and cellular localization patterns of anthocyanin accumulation

The growth of all plants was morphologically similar in the growth chamber except for color difference (Fig. 2 a). Red pigmentation patterns were examined with a microscope. Trichomes, epidermis, and hypodermis of *PAP1* transgenic stems highly accumulated anthocyanins (Figs. 2 b and 3). Parenchyma cells around vascular bundle and particularly near phloem were clearly red or pink resulting from high accumulation of anthocyanins (Fig. 3 b, c and e). In addition, parenchymal cells around xylem were clearly red or pink due to the high accumulation of anthocyanins (Fig. 3 b, c and e). Cells in the pith and cortex were not observed to produce anthocyanins (Fig. 3 b, c and e). In contrast, all cells in WT stems did not produce anthocyanins in the same growth conditions (Fig. 2 a-b and Fig. 3 a, d and f). In addition, *ANR* transgenic plants did not produce anthocyanins either (Fig. 2 a and b). As reported previously for greenhouse-grown plants (Xie et al. 2006), *PAP1::ANR* transgenic plants showed obviously reduced levels of anthocyanins in plants (Fig. 2 a and b).

We previously reported the features of anthocyanin accumulation in *PAP1* transgenic leaves of greenhouse-grown plants. Anthocyanins were mainly localized in trichomes, epidermal cells, hypodermal cells, and parenchyma cells in veins (Xie et al. 2006). In this experiment, the accumulation patterns of anthocyanin in leaves were the same as ones in leaves of greenhouse-grown plants (Xie et al. 2006).

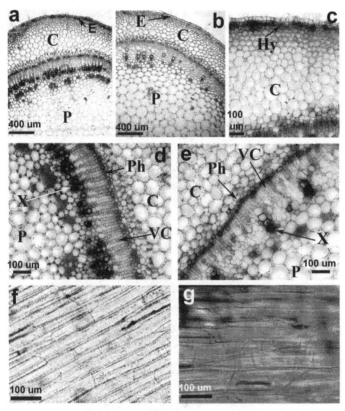

Fig. 3. Microscopic images show cellular specificity of anthocyanin accumulation in stems of
PAP1 transgenic plants. a-b cross sections of young stems of wild-type (a) and *PAP1*
transgenic (b) plants; c a magnified image showing anthocyanin accumulation in
hypodermal cells of *PAP1* transgenic stems; d a magnified image showing vascular bundle
of WT stems; e a magnified image showing anthocyanin accumulation patterns in or around
vascular bundles of *PAP1* transgenic stems; f epidermal cells of WT stems; g epidermal cells
of *PAP1* transgenic stems. Abbreviations: C, cortex; E, epidermis; Hy: hypodermis; P: pith;
Ph: phloem; VC: vascular cambium; X: xylem.

3.2 Metabolites identified and their profile properties in wild-type tobacco

Young leaves, old leaves, and stem tissues of WT tobacco plants were used to analyze
metabolites and examine their accumulation patterns. In total, eighty-seven metabolites
including both water (polar molecular compounds) and chloroform soluble compounds
(low or non-polar molecular compounds) were characterized based on their mass spectra
profile identity and chromatographic retention times in comparison to authentic standards.
ANOVA significance tests indicated that seventy-one metabolites were featured with tissue-
related accumulation patterns, which was shown in a heatmap (Fig. 4). Fourteen metabolites
did not show significant differences among the three tissues examined. In addition, two
metabolites, ribose and trihydroxybutyrate showed an interesting pattern in the three

tissues. ANOVA analysis showed that the levels of the two compounds were not obviously different between young leaves and stems as well as between old leaves and stems but significantly higher in young leaves than in old ones.

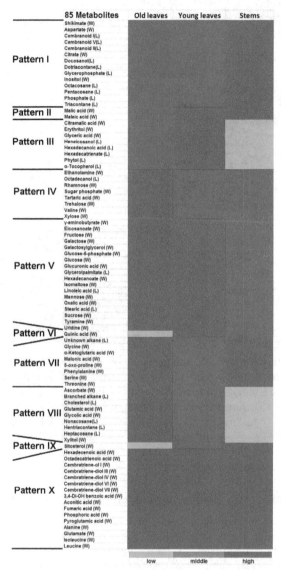

Fig. 4. A heatmap shows differential accumulation patterns of 85 metabolites in stems, young leaves, and old leaves. In each pattern, red color means the highest levels of metabolites in the tissue (P< 0.05), and blue color means significantly (P<0.05) higher levels of metabolites in the tissue than in the other tissue indicated by green colors. (L): lipophilic metabolites extracted in chloroform phase; (W): water soluble metabolites in extracted in water phase.

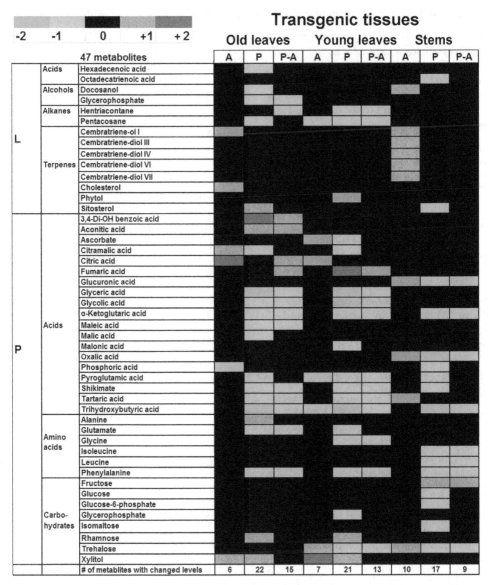

Fig. 5 A heatmap shows significant increase or decrease of levels for metabolite in *PAP1*,
ANR, and *PAP1::ANR* transgenic stems and young and old leaves. Black color (0) indicates
no differences in levels of metabolites between transgenic tissues and wild-type tissues. Red
(+2) color indicates highly significant (P<0.05) increase. Bright brown color (+1) indicates
significant (P<0.05) increase. Blue color (-1) indicates significant (P< 0.01) decrease. Green
color (-2) indicates highly significant (P<0.05) decrease. A: *ANR*; P: *PAP1*; P-A: *PAP1:: ANR*;
L: lipophilic metabolites extracted in chloroform phase; W: water soluble metabolites in
extracted in water phase.

3.3 Metabolites with significantly altered levels in *PAP1* transgenic plants

PAP1 is a master transcriptional regulator of the anthocyanin pathway (Fig. 1) (Xie et al. 2006). *PAP1* transgenic plants highly produced anthocyanins in the growth chamber (Fig. 2 a). GC-MS analysis of the eighty-seven metabolites showed that the levels of thirty-nine were altered, which included nine metabolites with increased levels, twenty-nine metabolites with reduced levels, and one metabolite, sitosterol, the level of which was increased in old leaf but decreased in stem (Fig. 5).

Phenylalanine, malonic acid and shikimic acid, which are three early pathway precursors of the anthocyanin biosynthesis (Fig. 1), were dramatically reduced (Fig. 5). The decreases of phenylalanine and shikimic acid were apparent in all three analyzed tissues, while malonic acid was significantly reduced only in young leaves (Fig. 5).

Thirty-six metabolites with altered levels were not immediately metabolically related to plant phenylpropanoids (Fig. 5). These metabolites included two clusters consisting of seven non-polar and twenty-nine polar compounds either in old leaves, young leaves, or stems. The seven non-polar metabolites, which were composed of acids, alkanes, alcohols, and terpenes, included 4 in old leaf, 3 in young leaf, and 2 in stem. For example, the level of non-polar compound docosanol was increased only in old leaf, while the abundance of non-polar compound sitosterol was reduced in stem but increased in old leaf. The twenty-nine polar metabolites, which were composed of acids, amino acids, and carbohydrates, included 18 in old leaf, 18 in young leaf, and 15 in stem. Examples of decreased accumulation included α-ketoglutaric acid, citramalic acid, glyceric acid, glycine, maleic acid, and trihydroxybutyric acid in a tissue specific fashion.

3.4 Metabolites with altered levels in *ANR* transgenic plants

Our previous work discovered that ANR (anthocyanidin reductase) was a pathway enzyme of the PA biosynthesis (Xie et al. 2003) . Analyses of variance followed by Tukey's HSD post-hoc comparisons were conducted to assess the alteration extent of metabolites in *ANR* transgenic plants. The levels of nineteen identified metabolites including eleven polar and eight non-polar metabolites were altered in tissue-dependent accumulation patterns (Fig. 5). The eleven polar metabolites included 4 in old leaf, 6 in young leaf, and 4 in stem, the levels of eight of which were increased in either leaves or stems, while, the levels of three of which were decreased in *ANR* transgenic leaves. The eight non-polar metabolites included seven with increased levels but one, pentacosane (a lipophilic odd-chained alkane) with a decreased level (Fig. 5). Among the seven metabolites with increased levels, cembratriene-ol I, cembratriene-diol III, cembratriene-diol IV, cembratriene-diol VI, and cembratriene-diol VII are defensive compounds against aphids (Wang et al. 2001).

3.5 Metabolites with altered levels in *PAP1: ANR* transgenic plants

PAP1::ANR transgenic plants were the F1 hybrid progeny of *ANR* and *PAP1* transgenic plants. These plants formed both anthocyanins and PAs (Xie et al. 2006). The levels of twenty-four metabolites including three non-polar and twenty-one polar metabolites were altered (Fig. 5). Among them, twenty-three were included in those metabolites with altered levels in *PAP1* transgenic plants as described above (Figs. 5 and 6). Phenylalanine and shikimic acid are two early precursors of the anthocyanin and PA pathways (Fig. 1). As observed in *PAP1* transgenic plants described above, the accumulation level of phenylalanine was significantly reduced in all analyzed tissues. The accumulation level of shikimic acid was also dramatically reduced in leaf tissues. Twenty-two metabolites with altered levels were not directly related to plant flavonoid biosynthesis, twenty-one of which were included in those impacted by*PAP1*alone transgenic plants (Fig. 5). Therefore, these results supported the metabolite profile alteration in *PAP1* alone transgenic plants.

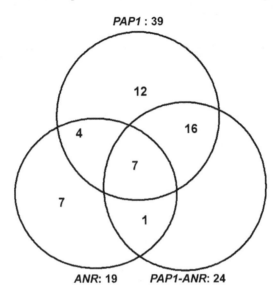

Fig. 6. A Venn diagram shows separate and overlaid properties of metabolites with altered levels in the *PAP1*, *ANR*, and *PAP1::ANR (PAP1-ANR)* transgenic plants. In total, the levels of forty-seven out of eighty-seven metabolites analyzed were impacted in the three transgenic events.

3.6 Principal component analysis

Principal component analysis (PCA) was performed with the eighty seven identified metabolites to examine the ordination relationships for all three tissues between WT plants and transgenic plants. The first principal component accounted for nearly 25%, 23% and 20% of total variation in old leaves, young leaves and stems respectively. The second principal component was nearly 12% of total variation in old leaves, young leaves, and stems. From PCA results for old and young leaves, it was obvious that *PAP1* transgenic plants and WT type plants were clearly separated each other in the first principal component (Fig. 7 a and b). *ANR* transgenic plants were scattered in the same cluster with WT plants. *PAP1::ANR* transgenic plants were between *PAP1* transgenic and *ANR* transgenic or WT plants (Fig. 7 a and b). It was interesting that in stems the ordination relationship patterns of four groups of plants were different from those observed in old and young leaves (Fig. 7 c). *PAP1* and *ANR* transgenic stems were of the most separation in the first principal component. WT and *PAP1* transgenic stems were also clearly different each other in the first principal component. As expected, *PAP1* and *PAP1::ANR* transgenic stems were relatively close (Fig. 7 c). These PCA results showed that the impacts of *PAP1* and *ANR* on metabolite profiles in transgenic plants were different.

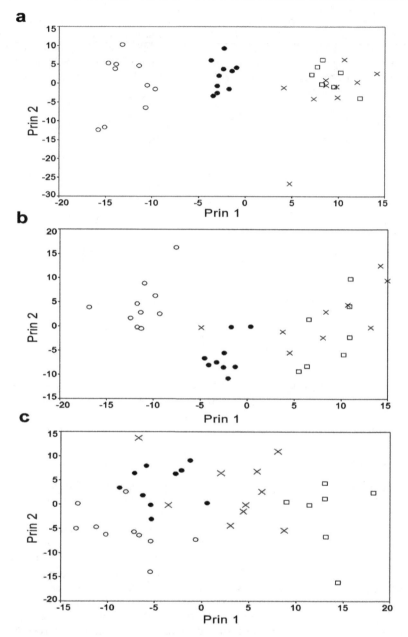

Fig. 7. Principal component analysis of wild-type plants ("x"), *ANR* (empty squares), *PAP1* (empty circles), or *PAP1::ANR* (solid circles) for (a) old leaf, (b) young leaf, and (c) stem tissue. Principal components 1 and 2 account for (a) 24.5 and 12.1 % of the total variation in old leaf, (b) 23.1 and 12.1% of the total variation in young leaf, and (c) 20.2 and 11.8% of the variation in stem tissue samples.

4. Discussion

4.1 Impacts of *PAP1* transgene on metabolic profiles

Our recent report showed that genome-wide transcriptional programs in red *pap1-D Arabidopsis* cells cultured *in vitro* were dramatically different from cultured wild-type cells (Shi and Xie 2011). On the one hand, massive numbers of genes involved in multiple metabolic networks were highly expressed in red cells. On the other hand, the expression levels of large number of genes involved in multiple metabolic networks were also reduced in red cells. We hypothesized that the dramatic alterations of genome-wide gene expression must have led to massive alterations of metabolic profiles beyond anthocyanin biosynthesis in red cells and specifically programmed red cells established by the overexpression of *PAP1* must have led to transcriptional and metabolic reprogramming (Shi and Xie 2011).

In this study, GC-MS analysis revealed that levels of thirty-nine metabolites excluding flavonoids were altered in *PAP1* transgenic plants (Figs 5 and 6). These metabolites are composed of seven non-polar molecules and thirty-two molecules including sugars, amino acids, and organic acids (Fig. 5). Based on their roles in anthocyanin biosynthesis, these metabolites can be characterized as targeted and non-targeted metabolites. Targeted metabolites are precursors in the anthocyanin pathway (Fig. 1). Phenylalanine, shikimic acid, and malonic acid are three early precursors of the anthocyanin biosynthesis. Therefore, the high production of anthocyanins must result from an enhanced metabolic flux from shikimic acid to anthocyanins (Fig. 1) and thus lead to the high consumption of shikimic acid, phenylalanine, and malonic acid. As expected, the levels of the three metabolites were reduced in *PAP1* transgenic plants. This result metabolically supported the previous reports about the effect of *PAP1* on gene expression profiles at the transcriptional level involved in anthocyanin biosynthesis. The transcriptional levels of the *PAL1* and several other anthocyanin pathway genes were increased 1.2-6.7 folds by the overexpression of the *PAP1* gene (Borevitz et al. 2000; Lillo et al. 2008; Tohge et al. 2005). In addition, this result supported our recent report that most of the pathway genes involved in the anthocyanin pathway in red cells over expressing *PAP1* were dramatically increased leading to alterations of multiple metabolic networks (Shi and Xie 2011). Therefore, the alterations of metabolic profiles of the three molecules resulted from the ectopic expression of *PAP1* transgene.

Most of metabolites with altered levels in *PAP1* transgenic tissues were not directly related to the biosynthesis of anthocyanins. They are non-targeted metabolites. These metabolites included acids, amino acids, and carbohydrates (Fig. 5). Two studies showed that the transcriptional levels of several genes involved in sugar transport, glutathione metabolism, calcium binding, and carbohydrate metabolism were up-regulated by PAP1 in *Arabidopsis* (Borevitz et al. 2000; Tohge et al. 2005). However, no changes in the levels of amino acids, sugars, or anions profiled by capillary electrophoresis or liquid chromatography were observed (Tohge et al. 2005). This result was not in agreement with our current observation. We observed dramatic reduction of levels of several amino acids and carbohydrates in response to *PAP1* expression in transgenic tobacco plants (Fig. 5). The discrepancy likely resulted from different experimental designs. We separated young leaves from old leaves in our experiments, while Tohge and co-workers (2005) did not in their study. More importantly, we recently isolated red cells from *pap1-D Arabidopsis*. Those red cells showed dramatically different transcriptional profiles from wild-type cells (Shi and Xie 2011) and these global gene expression alterations were not observed in Tohge's experiments either. Particularly, in red cells, expression levels of a large number of genes involved in glycolysis, amino metabolisms, photosynthesis and other metabolic pathways were significantly decreased (Shi and Xie 2011). In addition, the discrepancy might result from the species-specific responses to the over-expressed PAP1 protein. The results of ectopic expression of the *A. thaliana PAP1* gene in tobacco plants might differ from the consequences of its overexpression in its native plant species. The evidence was that we recently observed dramatically differential response of anthocyanin biosynthesis to chemical factors in tissue cultures of *PAP1* transgenic *Arabidopsis* and tobacco (Shi and Xie 2011; Zhou et al. 2008). Therefore, the alterations of these non-targeted metabolite profiles most likely resulted from the ectopic expression of *PAP1* transgene.

4.2 Impacts of ANR transgene on metabolic profiles

It was very interesting that our data showed the altered levels of nineteen metabolites in leaf and stem tissues of the *ANR* transgenic plants. Obviously, the accumulation of five cembranoids was enhanced in the stem of the *ANR* alone transgenic plants (Fig. 5). Cembranoids are diterpenoids derived from the isoprenoid pathway and involved in defense against aphids (Wang et al. 2001). In *Nicotiana spp.*, relatively high levels of cembranoids were found to localize to secretory glandular trichomes (Wang et al. 2004; Wang et al. 2001). In addition, oxalic acid accumulated at higher levels in *ANR* plants in a pattern similar to that of the cembranoids. Oxalate, which accumulates in glandular trichomes and idioblast cells of *Nicotiana* species (Choi et al. 2001; Sarret et al. 2006; VolkA and Franceschi 2000), is the calcium salt of oxalic acid and has important defensive activities against chewing insects (Korth et al. 2006). We microscopically examined trichomes, but did not observe changes of their morphology and density on the surfaces of stems and leaves between transgenic and WT tobacco plants (Fig. 2 b), suggesting that their level increase most likely resulted from the *ANR* transgene expression.

The mechanism behind the metabolic alteration in *ANR* transgenic plants is unclear. Little has been reported regarding the mechanism through which ectopic expression of a single biosynthetic reductase can dramatically alter cellular metabolome. Here, we would provide certain perspective discussion to potentially interpret this result observed from our experiments. A few studies revealed the effects of marker genes on *in vivo* cellular biochemical processes. Constitutive expression of marker genes such as *NPTII* and *GUS* genes resulted in changes of certain transcriptome profiles in transgenic *A. thaliana*, although there was no impact on the development and growth of plants (Ouakfaoui and Miki 2005). The constitutive presence of Bt protein was observed to unexpectedly increase the production of lignin in insect-resistant transgenic corn plants (Saxena and Stotzky 2001). However, the mechanism of such observed effects caused by the presence of introduced foreign proteins has not been investigated in detail thus remains unknown. It is interesting that a report demonstrated that an overexpression of a rice dihydroflavonol reductase (DFR) led to enhanced levels of NAD (H) and NADP (H) in rice cell suspension and seedlings, which increased the tolerance of transgenic rice to bacterial pathogens and reactive oxygen species (ROS)-induced cell death (Hayashi et al. 2005). DFR is a NADPH/NADH dependent reductase essential for both anthocyanin and PA biosynthesis (Fig. 1) (Xie et al. 2004a). Although Hayashi et al (2005) did not report whether or not their transgenic rice seedlings and cells had *in vivo* substrates for DFR enzyme, nor did they measure changes in unrelated metabolic pathways, the constitutive presence of DFR protein increased the levels of the two coenzymes by enhancing the enzymatic activities of NAD synthetase, NAD kinase and ATP-NMN adenylyltransferase (Hayashi et al. 2005). ANR is a DFR-like reductase using NADH/NADPH as coenzyme to convert anthocyanidins to two isomers of flavan-3-ols, e.g. (-)-epicatechin and (-)-catechin (Devic et al. 1999; Xie et al. 2003). *In vitro* experiments showed that ANR bound NADPH and NADH with Km values of 0.5 mM and 1.0 mM respectively (Xie et al. 2004b). In addition, our previous studies demonstrated the constitutive expression of the enzyme and its function to using *in vivo* NADPH/NADH as co-factor (Xie et al. 2006). The enzyme was recently localized to the cytoplasm as predicted (Pang et al. 2007). The concentration of cytosolic NADPH/NADH in tobacco plants is unknown, but a report showed that the cellular NADPH concentrations during photosynthesis were estimated to range from 0.1 to 0.5 mM in the blue-green algae *Anacystis nidulans* (Grossman and McGowan 1975). We hypothesize that the presence of ANR in transgenic tobacco plants may alter metabolic homeostasis of NAD or NADP metabolism as the observed DFR expression in rice (Hayashi et al. 2005). Thus, metabolism may be impacted by the changes of cellular homeostasis of NADPH/NADH. For example, sugar alcohol biosynthesis in celery is thought to be regulated by cytosolic NADPH levels (Gao and Loescher 2000). Consequently, the alterations of tissue-dependent patterns of numerous compounds observed in our experiment, e.g. trichomes-secreted cembranoids, likely result from the changes of coenzyme homeostasis or other factors.

4.3 Impacts of the coupled expression of *PAP1* and *ANR* transgenes on metabolic profiles

The twenty-four metabolite profiles altered in the *PAP1::ANR* transgenic F1 progeny plants were also impacted in *ANR* or *PAP1* transgenic plants (Figs. 5 and 6). It was interesting that obviously overlaid properties of metabolic profiles were observed between *PAP1* or *ANR* and *PAP1::ANR* plants (Fig. 5, Fig. 6). Without considering the effects of tissue specificity, it was observed that the accumulation levels of eleven identical metabolites, i.e. pentacosane , ascorbic acid, citramalic acid, glucuronic acid, oxalic acid, pyroglutamic acid, tartaric acid, trehalose, trihydroxybutyric acid, and xylitol, which were altered in *PAP1::ANR* transgenic plants, were increased or decreased in *PAP1* or/and *ANR* transgenic plants (Figs. 5 and 6). Twenty three of the twenty-four metabolites, which were showed their accumulation pattern alterations in *PAP1::ANR* transgenic plants, were also with altered levels in *PAP1* transgenic plants (Figs. 5 and 6).

It was interesting that a counteracting consequence on metabolite accumulation patterns was observed in *PAP1::ANR* transgenic plants due to the co-expression of *PAP1* and *ANR*. On the one hand, metabolite accumulation patterns altered in *ANR* transgenic plants were counteracted by the expression of *PAP1* in *PAP1::ANR* transgenic plants. Eleven metabolites with altered levels in *ANR* transgenic plants were not found to have the alteration in *PAP1::ANR* transgenic plants. For example, the levels of cembratriene-diol III, IV, VI and VII were increased in *ANR* transgenic stems, but in *PAP1::ANR* transgenic stems, reduced to similar levels as the ones in WT plants (Fig. 5). In addition, the levels of cembratriene-ol I, cholesterol, citramalic acid, and xylitol were increased in old leaves of the *ANR* transgenic plants but were reduced in the *PAP1::ANR* transgenic plants (Fig. 5). Other examples include citric acid, oxalic acid, ascorbic acid, and citramalic acid (Fig. 5). On the other hand, metabolite accumulation alterations caused by the expression of *PAP1* were counteracted by the expression of *ANR*. Levels of thirty-nine metabolites were altered in *PAP1* transgenic plants, among which the alteration of the levels of fifteen metabolites was diminished in the *PAP1::ANR* transgenic plants (Fig.5). For example, the levels of glucose-6-phosphate and glucose were reduced in *PAP1* transgenic stems, but in *PAP1* × *ANR* transgenic stems, returned to the normal levels as the ones in WT plants (Fig. 5). No evidence has been reported describing counteracting impacts of the co-expression of a regulatory gene and a structure gene from the flavonoid pathway on cellular metabolomes. We suggest that this phenomenon results from the reduction of transgene dosage in *PAP1::ANR* F1 progeny plants. Given that *PAP1* transgenic plants were homozygotic plants, the transgene dosage was higher in the mother plants than *PAP1::ANR* F1 progeny plants. In addition, the mechanism of counteracting phenomenon is likely more complicated than the additive of the two transgenes. This may be also due to the metabolon formation (metabolic channeling) consisting of multiple proteins required for every single secondary metabolite and for avoiding metabolic interference (Jorgensen et al. 2005; Winkel-Shirley 1999).

5. Conclusion

The *PAP1* gene encodes a R2R3-MYB transcription factor and its ectopic expression led to the formation of a complete anthocyanin pathway in all tissues of tobacco (*Nicotiana tobacum*) plants. Anthocyanidin reductase (ANR) catalyzes anthocyanidins to flavan-3-ols and its ectopic expression of *ANR* formed an incomplete pathway of flavan-3-ols and proanthocyanidins (PAs) in leaves and stems of tobacco plants. The co-expression of *PAP1* and *ANR* constructed a complete biosynthetic pathway of PAs in tobacco leaves, stems and flowers. Gas chromatography-mass spectrometry (GC-MS) analysis identified eighty-seven metabolites in wild-type tobacco leaves and stems, and revealed tissue-specific and development-dependent patterns of the abundance of seventy-one metabolites in plants. In *PAP1* transgenic plants, the accumulation patterns of thirty-nine metabolites (out of eighty-seven) including both primary and secondary metabolites were altered in either leaves or stems. In *PAP1::ANR* transgenic plants, twenty-four metabolites showed accumulation pattern alterations. Shikimic acid, phenylalanine, and malonic acid, which are three early precursors of the flavonoid pathway, were altered in their abundance in these two transgenic events. In ANR transgenic plants, nineteen metabolites showed abundance alterations. The array of metabolites altered in ANR transgenic plants was distinct from that in *PAP1* transgenic plants. The array of metabolites altered in *PAP1::ANR* transgenic plants were intermediate to the two transgenic events. Principal component analysis showed that there was close ordination relevance in metabolic profile alterations between *PAP1* alone and *PAP1::ANR* transgenic plants but distant ordination relevance between *PAP1* and *ANR* transgenic plants or wild-type plants. These results indicated that the ectopic expression of *PAP1*, *ANR*, or *PAP1::ANR* leading to different branch pathways of plant flavonoids differentially impacted accumulation patterns of metabolites.

6. Acknowledgments

The completion of this work benefited from the support of the Noble Foundation and the support of an USDA grant (USDA 2006-35318-17431). We appreciate constructive advice provided by Lloyd Sumner and Richard Dixon. We appreciate Dr. Bettina E. Broeckling for her critical reading and suggestions.

7. References

Borevitz JO, Xia Y, Blount J, Dixon RA, Lamb C (2000) Activation tagging identifies a conserved MYB regulator of phenylpropanoid biosynthesis. Plant Cell 12: 2383-2394

Broeckling CD, Huhman DV, Farag MA, Smith JT, May GD, Mendes P, Dixon RA, Sumner LW (2005) Metabolic profiling of Medicago truncatula cell cultures reveals the effects of biotic and abiotic elicitors on metabolism. J. Exp. Bot. 56: 323-336

Broeckling CD, Reddy IR, Duran AL, Zhao X, Sumner LW (2006) MET-IDEA: Data extraction tool for mass spectrometry-based metabolomics. Anal. Chem. 78: 4334-4341

Choi YE, Harada E, Wada M, Tsuboi H, Morita Y, Kusano T, Sano H (2001) Detoxification of cadmium in tobacco plants: formation and active excretion of crystals containing cadmium and calcium through trichomes. Planta 213: 45-50

Devic M, Guilleminot J, Debeaujon I, Bechtold N, Bensaude E, Koornneef M, Pelletier G, Delseny M (1999) The BANYULS gene encodes a DFR-like protein and is a marker of early seed coat development. Plant J 19: 387-398

Gao Z, Loescher WH (2000) NADPH supply and mannitol biosynthesis. Characterization, cloning, and regulation of the non-reversible glyceraldehyde-3-phosphate dehydrogenase in celery leaves. Plant Physiol. 124: 321-330

Gonzalez A, Mendenhall J, Huo Y, Lloyd A (2009) TTG1 complex MYBs, MYB5 and TT2, control outer seed coat differentiation. Develop Bio 325: 412-421

Grossman A, McGowan RE (1975) Regulation of glucose 6-phosphate dehydrogenase in blue-green algae Plant Physiol. 55: 658-662

Halket J, Przyborowska A, Stein S, Mallard W, Down S, Chalmers R, . . . , . (1999) Deconvolution gas chromatography/mass spectrometry of urinary organic acids--potential for pattern recognition and automated identification of metabolic disorders. Rapid Commun Mass Spectrom 13: 279-284

Hayashi M, Takahashi H, Tamura K, Huang J, Yu L-H, Kawai-Yamada M, Tezuka T, Uchimiya H (2005) Enhanced dihydroflavonol-4-reductase activity and NAD homeostasis leading to cell death tolerance in transgenic rice. PNAS 102: 7020-7025

Jorgensen K, Rasmussen AV, Morant M, Nielsen AH, Bjarnholt N, Zagrobelny M, Bak S, Moller BL (2005) Metabolon formation and metabolic channeling in the biosynthesis of plant natural products. Cur. Opin.Plant Biol. 8: 280-291

Korth KL, Doege SJ, Park S-H, Goggin FL, Wang Q, Gomez SK, Liu G, Jia L, Nakata PA (2006) Medicago truncatula mutants demonstrate the role of plant calcium oxalate crystals as an effective defense against chewing insects. Plant Physiol. 141: 188-195

Lillo C, Lea US, Ruoff P (2008) Nutrient depletion as a key factor for manipulating gene expression and product formation in different branches of the flavonoid pathway. Plant Cell Environ 31: 587-601

Ouakfaoui SE, Miki B (2005) The stability of the Arabidopsis transcriptome in transgenic plants expressing the marker genes nptII and uidA. Plant J. 41: 791-800

Pang Y, Peel GJ, Wright E, Wang Z, Dixon RA (2007) Early steps in proanthocyanidin biosynthesis in the model legume Medicago truncatula. Plant Physiol 145: 601-615

Ramsay NA, Glover BJ (2005) MYB-bHLH-WD40 protein complex and the evolution of cellular diversity. Trends Plant Sci 10: 63-70

Rowan DD, Cao M, Kui L-W, Cooney JM, Jensen DJ, Austin PT, Hunt MB, Norling C, Hellens RP, Schaffer RJ, Allan AC (2009) Environmental regulation of leaf colour in red 35S:PAP1 *Arabidopsis thaliana*. New Phytologist 182: 102-115

Sarret G, Harada E, Choi Y-E, Isaure M-P, Geoffroy N, Fakra S, Marcus MA, Birschwilks M, Clemens S, Manceau A (2006) Trichomes of tobacco excrete zinc as zinc-substituted calcium carbonate and other zinc-containing compounds. Plant Physiol. 141: 1021-1034

Saxena D, Stotzky G (2001) Bt corn has a higher lignin content than non-Bt corn. Am. J. Bot. 88: 1704-1706

Shi M-Z, Xie D-Y (2010) Features of anthocyanin biosynthesis in *pap1-D* and wild-type *Arabidopsis* thaliana plants grown in different light intensity and culture media conditions. Planta 231: 1385-1400

Shi M-Z, Xie D-Y (2011) Engineering of red cells of *Arabidopsis thaliana* and comparative genome-wide gene expression analysis of red cells versus wild-type cells Planta DOI: 10.1007/s00425-010-1335-2: In press

Springob K, Nakajima H, Yamazaki M, Saito K (2003) Recent advances in the biosynthesis and accumulation of anthocyanins. Nat. Prod. Rep. 20: 288-303

Tanner GJ, Francki KT, Abrahams S, Watson JM, Larkin PJ, Ashton AR (2003) Proanthocyanidin biosynthesis in plants. Purification of legume leucoanthocyanidin reductase and molecular cloning of its cDNA. J Biol Chem 278: 31647-31656

Tohge T, Nishiyama Y, Hirai MY, Yano M, Nakajima J, Awazuhara M, Inoue E, Takahashi H, Goodenowe DB, Kitayama M, Noji M, Yamazaki M, Saito K (2005) Functional genomics by integrated analysis of metabolome and transcriptome of *Arabidopsis* plants over-expressing an MYB transcription factor. Plant J 42: 218-235.

VolkA GM, Franceschi VR (2000) Localization of a calcium channel-like protein in the sieve element plasma membrane. Aust. J. Plant Physiol. 27: 779-786

Wang E, Hall JT, Wagner GJ (2004) Transgenic *Nicotiana tabacum* L. with enhanced trichome exudate cembratrieneols has reduced aphid infestation in the field. Mol. Breeding 13: 49-57

Wang E, Wang R, DeParasis J, Loughrin JH, Gan S, Wagner GJ (2001) Suppression of a P450 hydroxylase gene in plant trichome glands enhances natural-product-based aphid resistance. Nat Biotech 19: 371-374

Winkel-Shirley B (1999) Evidence for enzyme complexes in the phenylpropanoid and flavonoid pathways. Physiol. Plant. 107: 142-149

Xie D-Y, Jackson LA, Cooper JD, Ferreira D, Paiva NL (2004a) Molecular and biochemical analysis of two cDNA clones encoding dihydroflavonol-4-reductase from *Medicago truncatula*. Plant Physiol. 134: 979-994

Xie D-Y, Sharma SB, Dixon RA (2004b) Anthocyanidin reductases from *Medicago truncatula* and *Arabidopsis thaliana*. Arch Biochem Biophys 422: 91-102

Xie D-Y, Sharma SB, Paiva NL, Ferreira D, Dixon RA (2003) Role of anthocyanidin reductase, encoded by *BANYULS* in plant flavonoid biosynthesis. Science 299: 396-399

Xie D-Y, Sharma SB, Wright E, Wang Z-Y, Dixon RA (2006) Metabolic engineering of proanthocyanidins through co-expression of anthocyanidin reductase and the PAP1 MYB transcription factor. Plant J 45: 895-907

Zhou L-L, Zeng H-N, Shi M-Z, Xie D-Y (2008) Development of tobacco callus cultures over expressing *Arabidopsis* PAP1/MYB75 transcription factor and characterization of anthocyanin biosynthesis Planta 229: 37-51

7

Effect of Antisense Squalene Synthase Gene Expression on the Increase of Artemisinin Content in *Artemisia anuua*

Hong Wang et al.*
Graduate University of the Chinese Academy of Sciences, Beijing, China

1. Introduction

Artemisinin, a sesquiterpene lactone endoperoxide, is a valuable and powerful antimalarial drug obtained from the aerial parts of a Chinese herb, *Artemisia annua* (Liu et al., 1979). Artemisinin and its derivatives show few or no side effects with the existing antimalarial drugs, it has consequently been regarded as the next generation of antimalarial drugs (Looaresuwan, 1994). Currently, commercial production of artemisinin mainly based on its extraction and purification from plant material, however, the endogenous production of artemisinin is very low (0.01%-0.8% dry weight) (Wallaart et al., 1999). In view of the limited availability of artemisinin and the increased demand, the synthetic preparation of artemisinin becomes an attractive proposition. However due to its complex structure, the complete chemical synthesis is very difficult (Schmid & Hofheinz, 1983). *Artemisia annua* as the only valid source, many research groups have directed their investigations toward the enhancement of artemisinin production in *A. annua* cell cultures or whole plants by biotechnological approaches. However these approaches were still proved to be not successful (Ghingra et al., 2000). Recently, several genes in artemisinin biosynthesis have been cloned, and important advances in artemisinin biosynthesis have been achieved, which makes it possible to regulate artemisinin biosynthesis in a direct way, for example, by metabolic engineering (Abdin et al., 2003).

Artemisinin is synthesized by the isoprenoid pathway. In the cytosol, isoprenoids are synthesized via the classical acetate/mevalonate pathway (Fig. 1). In this pathway, farnesyl diphosphate (FDP) occupies a central position and serves as a common substrate for the first committed reactions of sterols and sesquiterpenes, such as artemisinin. Therefore, this point represents a potentially important controlling point for balancing sterol synthesis and sesquiterpenes synthesis. From metabolic engineering point of view (Fig. 1), there are two

*Yugang Song[1], Haiyan Shen[1], Yan Liu[2], Zhenqiu Li[3], Huahong Wang[2], Jianlin Chen[2], Benye Liu[2] and Hechun Ye[2]
[1]Graduate University of the Chinese Academy of Sciences, Beijing, China
[2]Key Laboratory of Photosynthesis and Environmental Molecular Physiology, Institute of Botany, the Chinese Academy of Sciences, Beijing, China
[3]Hebei University, Baoding, China*

ways to increase the flux to artemisinin biosynthesis, on the one hand, we can overexpress the key genes involved in the biosynthesis of artemisinin; on the other hand, we can inhibit the genes involved in other pathways competing for its precursors.

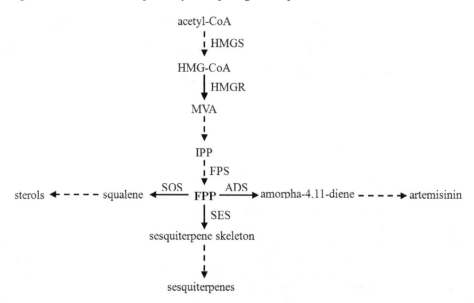

ADS: amorpha-4,11-diene synthase; FPP: farnesyl diphosphate; FPS: farnesyl diphosphate synthase; HMG-CoA: 3-hydroxy-3-methylglutaryl CoA; HMGR: HMG-CoA reductase; HMGS: HMG-CoA synthase; IPP: isopentenyl diphosphate; MVA: mevalonate; SES: sesquiterpene synthase; SQS: squalene synthase.

Fig. 1. Diagram of the mevalonate pathway leading to the biosynthesis of sesquiterpenes and sterols.

Squalene synthase (SQS) catalyzes the condensation of two molecules of farnesyl diphosphate (FDP) to form the linear 30 carbon compound squalene, the first committed precursor for sterol biosynthesis (Goldstein & Brown, 1990). SQS is generally described as a crucial branch point enzyme for synthesizing sterol intriguing as a potential regulatory point that controls carbon flux into either sterol or into non-sterol isoprenoids (such as sesquiterpenes). So if the SQS gene expression is inhibited by genetic manipulation, the carbon flux into sterol may be diverted to sesquiterpenes, and the biosynthesis of sesquiterpenes may be increased. With the purpose to increase artemisinin production, we have cloned squalene synthase cDNA (SQS) (Liu et al., 2003). In this chapter, we report the construction of the antisense SQS plant expression vector, and its effects on inhibition of SQS gene expression on squalene and artemisinin biosynthesis.

2. Materials and methods

2.1 Plant materials

A high artemisinin producing *Artemisia annua* L. strain 001 was collected from Sichuan Province of China. The seeds were surface sterilized and cultured on the Murashige &

Skoog (1962) basal medium with 0.7% agar and 3% sucrose in growth chamber at 26 °C and 16 h photoperiods. Leaves of 2-week-old seedlings were used for *Agrobacterium*-mediated transformation. After transformation, the transgenic plants and control plants were grown in green house with natural light and watered manually.

2.2 Construction of antisense plant gene expression vector

The original plasmid pSQF2 containing squalene synthase cDNA (SQS) gene, cloned from *A. annua* by our laboratory (Liu et al., 2003), was used as a template for the amplification of SQS fragment, then the amplified fragment was used for the construction of antisense plant expression vector. The PCR primers were designed according to the sequence at the 5'- and 3'-terminal region of SQS gene: 5'-GAC GGA TCC AAC AAA CAG TAC AAT TGG TG-3' (*Bam*H I restriction site underlined) and 5'-GCA GAG CTC GGA TTT GGA TCT TGA AGA AG-3' (*Sac* I restriction site underlined). The amplified SQS fragment was 1.5 kb in length, after digested with *Bam*H I and *Sac* I, the resulting large fragment was collected. The binary vector pBI121, containing both the *NPT II* gene controlled by the NOS promoter and the *GUS* gene controlled by the cauli-flower mosaic virus 35S promoter, was digested with *Bam*H I and *Sac* I, and the resulting large fragment was collected. The above two collected fragments were fused with T4 DNA ligase (Takara) (Fig. 2). The recombinant plasmid pBISQS was transformed into the competent *E. coli* DH5α prepared using CaCl$_2$ method and pBISQS was extracted using alkaline lysis method then sequenced by Genecore Company.

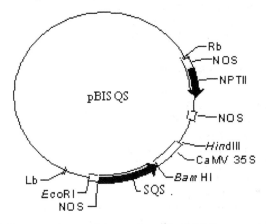

Fig. 2. Structure and diagram of recombinant plasmid pBISQS.

2.3 *Agrobacterium* preparation and genetic transformation

The binary vector pBISQS was introduced into *Agrobacterium tumefaciens* strain EHA105 by the freeze-thaw method and used for genetic transformation. Transformants were selected on LB medium supplemented with 50 mg/L kanamycin (Kan) and 50 mg/L rifampicin. The corresponding wild type strain was cultured in LB medium containing 50 mg/L rifampicin.

Plant transformation was basically performed according to the method of Han et al (2005). The leaves of 2-week-old aseptic seedlings were immersed in 50 mL *A. tumefaciens*, which was at the log phase of growth and was diluted 10-fold with MS medium. After 20 min, the infected leaves were taken out, blotted with sterile paper and co-cultured on solid MS medium at 26 °C for 2 or 3 d. After co-cultivation, the leaves were transferred to shoot-inducing medium (MS medium supplemented 1.0 mg/L 6-BA and 0.05 mg/L NAA) containing 20 mg/L Kan to induce shoot and 500 mg/L cefotaxine (Cefo) to kill residual *Agrobacterium*. The medium without Kan was used as a control. Here the key difference with Han's method is that a sheet of sterile filter paper was placed on the shoot-inducing medium, since we found by such a simple action, the shoot induction frequency can be noticeably increased. One week later, the leaves were transferred to fresh MS selection medium with 400 mg/L Cefo. After 4-6 weeks selection Kan resistant shoots were obtained and transferred to 100 mL Erlenmeyer flask with 40 mL regeneration medium (MS basal medium supplemented with 0.05 mg/L NAA) to induce roots. All the rooted seedlings were kept in the greenhouse at 25 °C under the fluorescent lamps (with a light intensity of 3000 lx, 16h), with a relative humidity of 40%.

2.4 PCR detection

The integration of SQS gene into *A. annua* genome was confirmed by the polymerase chain reaction (PCR). The CTAB method was used to purify the genomic DNA from transgenic *A. annua* leaves. The forward primer was 5'-CCA CGT CTT CAA AGC AAG TGG ATT -3', designed according to the sequence of CaMV 35S promoter, and the reverse primer was 5'-GCA GAG CTC GGA TTT GGA TCT TGA AGA A G -3', designed according to the sequence at the 3'-terminal region of SQS gene. 30-cycle reactions, each consisted of heated denaturation (94 °C for 40 s), annealing (55 °C for 30 s) and extension (72 °C for 2 min), were carried out and the reaction mixtures were subjected to agarose gel electrophoresis.

2.5 RNA isolation and RT-PCR detection

For reverse transcription-polymerase chain reactions (RT-PCR), 1 g of total RNA isolated from the leaves of the transformed and non-transformed plants was used. The RNA extraction protocol was done as described in Sambrook et al (1989). The first strand cDNA was synthesized by using a first-strand cDNA synthesis kit (TaKaRa), according to the manufacturer's instructions. The resultant first-strand cDNA was used as a template, and the PCR primers, P1: 5'-GGA ACC ATG GGT AGT TTG AAA GCA GTA TTG-3, and P2: 5'-GCC TGG ATC CCT TGA CTC TCT CTT AAC TAT-3', were designed according to the SQS gene sequence of *A. annua*. The PCR was performed in the same way as described in the Section of PCR detection. At the same time, to normalize the amount of mRNA in each PCR reaction, a PCR product of actin in *A. annua* was amplified, and the primers were: 5'-AAC TGG GAT GAC ATG GAG AAG ATA T-3', and 5'-TCA CAC TTC ATG ATG GAG TTG TAG G-3'.

2.6 Squalene analysis

Squalene analysis was carried out according to the method described by Wentzinger et al (2002). For all the plants analyzed, young leaves was collected. Extraction and purification of the samples were in accordance with Wentzinger et al (2002), and the samples were analyzed by GC, the GC injection port was operated at 120 °C. The oven temperature programmed from 120 °C to 180 °C at 15 °C min^{-1} and from 180 °C to 260 °C at rate of 25 °C min^{-1}. The final temperature was maintained for 25 min. The results were compared to standards.

2.7 Determination of artemisinin

The detection of artemisinin was performed according to the method of Zhao & Zeng (1986). Fresh leaves of the transformed and non-transformed plants were collected and dried to constant weight in an oven at 50 °C. Then the dried leaves were ground to fine powder. Exactly 0.05 g powder was added to an extraction bottle containing 40 mL petroleum ether (30-60 °C) and treated in a supersonic bath for 2 min. The extraction mixture was filtered and the petroleum ether was evaporated. The residue was dissolved in 1 mL methanol and centrifuged at 12000 r/min to precipitate the undissolved components. The supernatant was used for detection of artemisinin by HPLC.

200 µL of the above prepared methanol solution was placed in a 10 mL tube and 800 µL methanol and 4 mL 0.2% sodium hydroxide were added, mixed and maintained in a 50 °C water bath for 30 min, then the reaction mixture was cooled to room temperature. 0.5 mL of the reaction mixture was placed in a 1.5 mL Eppendorf tube, and 100 µL methanol and 400µL 0.05 M acetic acid were added, mixed and the sample purified by filtering on a NC filter (40 µm). The artemisinin standard (Sigma, MO) solutions with concentrations of 3, 6, 12, 24 and 48 µg/mL were prepared in the same way as the sample.

C18 reverse column was 4.6×250 mm, 5 µm. The mobile phase was 0.01 M phosphate buffer (pH 7.0) : methanol (55:45), with flow rate 1 mL/min. The wavelength of the UV detector was 288.6 nm and the injection volume was 20 µL. Artemisinin standard appeared at 4 min 30 s under the above mentioned conditions.

3. Results

3.1 Regeneration of transgenic *A. annua*

Leaf discs, which were infected with *A. tumefaciens* strain EHA105 harboring the binary vector pBISQS, were co-cultured for 36-48 h at 26 °C in dark, then transferred to shoot-inducing medium. Shoots usually start to appear within 2-3 weeks on this medium. In order to obtain higher frequency of shoot induction, a sheet of filter paper was placed on the shoot-inducing medium during the transformant selection step, which was proved to be a very effective means for shoot induction (Song et al., 2006). Regenerated shoots are rooted on MS medium with 0.05 mg/L NAA, followed by transferring to greenhouse (16 h light at 25 °C).

3.2 Molecular analysis of transgenic plants

To investigate the presence of antisense SQS gene in the putatively transformed plants, genomic DNA of 4 Kan-resistant plants regenerated from the leaves inoculated with EHA105 was isolated, and PCR analysis was performed. The antisense SQS gene was detected as 1660-bp fragments in all 4 analyzed plants (Fig. 3). The amplified fragments were of the same size as the predicted one. The fragment in the nontransformed plant was not amplified.

RT-PCR was performed using specifically designed primers according to the squalene synthase cDNA sequence in *Artemisia annua*. These primers allow specific amplification of *A. annua* squalene synthase cDNA. The results showed that the suppressed expression of the endogenous *A. annua* squalene synthase gene in lines SQS3 and SQS5, but the transcriptional level in line SQS2 had no noticeable difference to that of the control (Fig. 4).

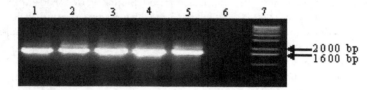

1. positive control; 2-5. transformed lines; 6. negative control; 7. DNA ladder.

Fig. 3. PCR amplification of the transformed plants and the control.

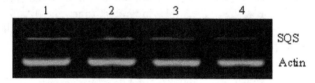

1. 001 line (non-transgenic control); 2. SQS2; 3. SQS3; 4. SQS5.

Fig. 4. RT-PCR analysis of the transformed plants and the control.

3.3 Detection of squalene content

In order to determine the effects of inhibiting squalene synthase gene expression on sterol biosynthesis, the leaves of transgenic lines SQS3, SQS5, and these of the control 001 were selected to detect squalene content. The results of GC-MS showed that in SQS3 and SQS5 transgenic lines, squalene content is decreased by 19.4% and 21.6% respectively in comparison with the control (Fig. 5).

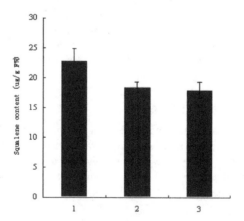

1. 001 line (control); 2. SQS3; 3. SQS5.

Fig. 5. Analysis of squalene content of the transgenic plants and the control.

3.4 Determination of artemisinin

Artemisinin was detected by HPLC. The results indicated that artemisinin content of SQS3 and SQS5 transgenic lines was increased by 23.2% and 21.5%, respectively compared with that of the control (Fig. 6) and in SQS2 transgenic line, the artemisinin content manifested no obvious variation compared with the control.

The above results demonstrated a clear negative correlation between squalene content and artemisinin content, which implies that the inhibiting of squalene synthase gene expression caused part of the flux for squalene biosynthesis diverting to artemisinin biosynthesis.

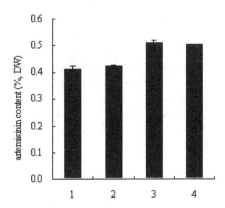

1. 001 line (control); 2. SQS2; 3. SQS3; 4. SQS5

Fig. 6. Analysis of artemisinin content of the transgenic plants and the control.

4. Discussion

Since squalene synthase is commonly depicted as the incipient and crucial branch point enzyme of the isoprenoid pathway to sterol biosynthesis, it has attracted considerable interest as a potential regulatory point that controls carbon flux into sterols. Several researchers reported the induction of sesquiterpene phytoalexins biosynthesis had been correlated with suppression of sterol biosynthesis in elicitor-treated tobacco cell cultures (Chappell et al., 1989; McGarvey & Croteau, 1995; Yin et al., 1997). The induction of one enzyme and suppression of the other are thought to be one mechanism that regulates the production of squalene and sesquiterpene (Devarenne et al., 1998). The biosynthesis of artemisinin belongs to the isoprenoid pathway, in this pathway, squalene synthase and amorpha-4,11-diene synthase are positioned at putative branch points in isoprenoid metabolism, these two enzymes catalyze the common farnesyl diphosphate to form squalene and amorph-4,11-diene, respectively. Furthermore, amorpha-4,11-diene synthase is considered as a key enzyme in artemisinin biosynthesis (Bouwmeester et al., 1999), so squalene synthase can be considered as a competitive enzyme of artemisinin biosynthesis. Therefore, the inhibiting of SQS gene expression may increase the biosynthesis of artemisinin.

In order to increase artemisinin content, we introduced antisense squalene synthase gene into *A. annua* via *Agrobacterium*-mediated transformation and transgenic plants were obtained. It has been shown that the transgenic plants had an increase of artemisinin content in lines SQS3 and SQS5 and a reduction in squalene content. This may be due to the suppression of squalene synthase gene expression caused part of the carbon flux to squalene biosynthesis diverting to artemisinin biosynthesis. At the same time, in correspondence with the decline in squalene content, the endogenous squalene synthase transcript level of the SQS3 and SQS5 transgenic lines were reduced compared with the control. These results strongly supported that the overexpression of squalene synthase in antisense orientation had a relevant effect on endogenous squalene metabolism in transgenic tobacco plants (Zhang et al., 2005). Though the SQS gene expression of the transgenic plants was inhibited, both the growth and the phenotype of the transgenic plants showed no obvious difference to those of the control, and this was in consistent with the results of transgenic tobacco (Zhang et al., 2005).

Our results as well as other related reports all indicated that it is possible to increase artemisinin content of *A. annua* by inhibiting the expression of genes competing precursors with artemisinin in transgenic plants (Yang et al., 2008; Zhang et al., 2009; Chen at el., 2011). But it has caused only a limited increase in artemisinin content, perhaps this is because of the biosynthesis of artemisinin is controlled by multi-genes and the contribution of manipulating one gene to artemisinin biosynthesis is limited. Recently, the study of artemisinin has made great progress, and more and more genes of artemisinin biosynthesis, such as cytochrome P450 monooxygenase (CYP71AV1), double bond reductase 2 (DBR2) and aldehyde dehydrogenase 1 (ALDH1) (Teoh et al., 2006; Covello et al., 2007; Zhang et al., 2008; Teoh et al., 2009), have been cloned, this makes it possible to regulate multi-genes of artemisinin biosynthesis in the future.

5. Conclusion

The antisense squalene synthase (SQS) gene was transferred into *A. annua* via *Agrobacterium*-mediated transformation, and the artemisinin content of one of the transgenic lines showed an increase of 23.2% in comparison to the wild-type control. The results demonstrated that inhibiting pathway competing for precursor of artemisinin by anti-sense technology is an effective means of increasing the artemisinin content of *A. annua* plants.

6. Acknowledgements

This work was supported by the National Natural Science Foundation of China (NSFC) (No. 61173098) and the Knowledge Innovation Program of the Chinese Academy of Sciences (KSCX2-EW-J-29).

7. References

Abdin, M., Israr, M., Rehman, R. & Jain, S. (2003). Artemisinin, a novel antimalarial drug: biochemical and molecular approaches for enhanced production. *Planta Medica*, Vol.69, No.4 (April 2003), pp. 289-299, ISSN 0032-0943

Bouwmeester, H., Wallaart, T., Janssen, M., van Loo, B., Jansen, B., Posthumus, M., Schmidt, C., De Kraker, J., Konig, W. & Franssen, M. (1999). Amorpha-4,11-diene synthase catalyses the first probable step in artemisinin biosynthesis. *Phytochemistry*, Vol.52, No.5 (November 1999), pp. 843-854, ISSN 0031-9422

Chappell, J., Von Lanken, C., Vögeli, U. & Bhatt, P. (1989). Sterol and sesquiterpenoid biosynthesis during a growth cycle of tobacco cell suspension cultures. *Plant Cell Reports*, Vol.8, No.1 (January 1989), pp. 48–52, ISSN 0721-7714

Chen, J., Fang, H., Ji, Y., Pu, G., Guo, Y., Huang, L., Du, Z., Liu, B., Ye, H., Li, G. & Wang, H. Artemisinin biosynthesis enhancement in transgenic *Artemisia annua* plants by downregulation of the β-caryophyllene synthase gene. *Planta Medica*, DOI: 10.1055/s-0030-1271038, Epub ahead of print, SSN 0032-0943

Covello, P., Teoh, K., Polichuk, D., Reed, D. & Nowak, G. (2007). Functional genomics and the biosynthesis of artemisinin. *Phytochemistry*, Vol.68, No.14 (July 2007), pp. 1864-1871, ISSN 0031-9422

Devarenne, T., Shin, D., Back, K., Yin, S. & Chappell, J. (1998). Molecular characterization of tobacco squalene synthase and regulation in response to fungal elicitor. *Archives of Biochemistry and Biophyics*, Vol.349, No.2 (January 1998), pp. 205–215, ISSN 0003-9861

Ghingra, V., Vishweshwar Rao, K. & Lakshmi Narasu, M. (2000). Current status of aremisinin and its derivatives as antimalarial drugs. *Life Sciences*, Vol.66, No.4 (December 1999), pp. 279-300, ISSN 0024-3205

Goldstein, J. & Brown, M. (1990). Regulation of the mevalonate pathway. *Nature*, Vol.343, No.6257 (February 1990), pp. 425–430, ISSN 0028-0836

Han, J., Wang, H., Ye, H., Liu, Y., Li, Z., Zhang, Y., Zhang, Y., Yan, F. & Li, G. (2005). High efficiency of genetic transformation and regeneration of *Artemisia annua* L. via *Agrobacterium tumefaciens*-mediated procedure. *Plant Science*, Vol.168, No.1 (January 2005), pp. 73-80, ISSN 0168-9452

Liu, J., Ni, M., Fan, J., Tu, Y., Wu, Z., Wu, Y. & Chou, W. (1979). Structure and reaction of arteannuin. *Acta Chimica Sinica*, Vol.37, No.2 (April 1979), pp. 129-143, ISSN0567-7351

Liu, Y., Ye, H., Wang, H. & Li, G. (2003). Molecular cloning, Escherichia coli expression and genomic organization of squalene synthase gene from *Artemisia annua*. *Acta Botanica Sinica*, Vol. 45, No.5 (May 2003), pp. 608-613, ISSN 0577-7496

Looaresuwan, S. (1994). Overview of clinical studies on artemisinin derivatives in Thailand. *Transactions of the Royal Society of Tropical Medicine and Hygiene*, Vol.88, No.suppl (June 1994), pp. s9-11, ISSN 0035-9203.

McGarvey, D., & Croteau, R. (1995). Terpenoid metabolism. *The Plant Cell*, Vol.7, No.7 (July 1995), pp. 1015–1026, ISSN 1040-4651

Murashige, T. & Skoog, F. (1962). A revised medium for rapid growth and bioassays with tobacco tissue cultures. *Physiologia Plantarum*, Vol.15, No.3 (September 1962), pp. 473-497, ISSN 0031-9317

Sambrook, J., Fritsch, E. & Maniatis, T. (1989). *Molecular cloning: A laboratory manual* (second edition), Cold Spring Harbor Laboratory Press, ISBN 0-87969-309-6, New York

Schmid, G. & Hofheinz, W. (1983). Total synthesis of qinghaosu. *Journal of the American Chemical Society*, Vol.105, No.3 (February 1983), pp. 624-625, ISSN 0002-7863

Song, Y., Li, Z., Wang, H., Liu, B., Li, G. & Ye, H. (2006). Effect of filter paper on shoot induction and genetic transformation of *Artemisia annua* L. *Chinese Journal of Applied and Environmental Biology*, Vol.12, No.6 (December 2006), pp. 788-791, ISSN 1006-687X

Teoh, K., Polichuk, D., Reed, D., Nowak, G. & Covello, P. (2006). *Artemisia annua* L. (Asteraceae) trichome-specific cDNAs reveal CYP71AV1, a cytochrome P450 with a key role in the biosynthesis of the antimalarial sesquiterpene lactone artemisinin. *FEBS Letters*, Vol.580, No.5 (February 2006), pp. 1411-1416, ISSN 0014-5793

Teoh, K., Polichuk, D., Reed, D. & Covello, P. (2009). Molecular cloning of an aldehyde dehydrogenase implicated in artemisinin biosynthesis in *Artemisia annua*. *Botany*, Vol.87, No.6 (June 2009), pp. 635–642, ISSN 1916-2790

Wallaart, T., Pras, N. & Quax, W. (1999). Seasonal variations of artemisinin and its biosynthetic precursors in tetraploid *Artemisia annua* plants compared with the diploid wild-type. *Planta Medica*, Vol.65, No.8 (December 1999), pp. 723-728, ISSN 0032-0943

Wentzinger, L., Bach, T. & Hartmann, M. (2002). Inhibition of squalene synthase and squalene epoxidase in tobacco cells triggers an up-regulation of 3-hydroxy-3-methylglutaryl coenzyme a reductase. *Plant Physiology*, Vol.130, No.1 (September 2002), pp. 334-346, ISSN 0032-0889

Yang, R., Feng, L., Yang, X., Yin, L., Xu, X. & Zeng, Q. (2008). Quantitative transcript profiling reveals down-regulation of a sterol pathway relevant gene and overexpression of artemisinin biogenetic genes in transgenic *Artemisia annua* plants. *Planta Medica*, Vol.74, No.12 (October 2008), pp. 1510–1516, ISSN 0032-0943

Yin, S., Mei, L., Newman, J., Back, K. & Chappell, J. (1997). Regulation of sesquiterpene cyclase gene expression. Characterization of an elicitor- and pathogen-inducible promoter. *Plant Physiology*, Vol.115, No.2 (October 1997), pp. 437–451, ISSN 0032-0889

Zhang, L., Jing, F., Li, F., Li, M., Wang, Y., Wang, G., Sun, X. & Tang, K. (2009). Development of transgenic *Artemisia annua* (Chinese wormwood) plants with an enhanced content of artemisinin, an effective anti-malarial drug, by hairpin-RNA-mediated gene silencing. *Biotechnology and Applied Biochemistry*, Vol.52, No.3 (March 2009), pp. 199–207, ISSN 0885-4513

Zhang, Y., Liu, Y., Wang, H., Ye, H. & Li, G. (2005). Regulation of squalene synthase gene expression in tobacco by antisense transformation with an *Artemisia annua* squalene synthase gene. *Journal of Agricultural Biotechnology*, Vol.13, No.4 (August 2005), pp. 416-422, ISSN 1674-7968

Zhang, Y., Teoh, K., Reed, D., Maes, L., Goossens, A., Olson, D., Ross, A. & Covello, P. (2008). The molecular cloning of artemisinic aldehyde Delta 11(13) reductase and its role in glandular trichome-dependent biosynthesis of artemisinin in *Artemisia annua*. *The Journal of Biological Chemistry*, Vol.283, No.31 (August 2008), pp. 21501-21508, ISSN 0021-9258

Zhao, S., & Zeng, M. (1986). Determination of qinghaosu in *Artemisia annua* L. by high performance liquid chromatography. *Chinese Journal of Pharmaceutical Analysis*, Vol.6, No.1 (January 1986), pp. 3-5, ISSN 0254-1793

Part 2

Biosafety

8

Biosafety and Detection of Genetically Modified Organisms

Juliano Lino Ferreira[1], Geraldo Magela de Almeida Cançado[1],
Aluízio Borém[2], Wellington Silva Gomes[2] and Tesfahun Alemu Setotaw[1]
[1]Plant Biotechnology Laboratory, Agricultural Research Agency of
Minas Gerais - EPAMIG, Caldas – MG,
[2]Department of Crop Science, Universidade Federal de Viçosa, Viçosa-MG,
Brazil

1. Introduction

Biosafety is a set of actions focused on preventing, minimizing and eliminating risks associated with research, production, teaching, use, technology development and services related to genetically modified organisms (GMOs) with the aims of protecting human and animal health and environmental preservation.

Transgenic organisms, or GMOs, are organisms in which genetic material has been altered by recombinant DNA technology. Biotechnology allows the insertion of one or more genes into the genome of an organism from a different organism or species (e.g., animals, plants, viruses, bacteria); the expression of the introduced gene results in a new feature in the phenotype of the modified organism. A shortened definition of genes is that they are DNA sequences that contain the necessary information to affect phenotypic expression in an organism, such as the shape of a seed or resistance to a specific pest. The information encoded by the gene is expressed through two principal steps: transcription, in which the coding region of the DNA is copied into single-stranded RNA; and translation, in which the amino acid sequence encoded by mRNA is assembled and translated into protein. Thus, for the creation of a GMO, it is necessary to introduce the gene responsible for a particular trait into the genome of the target organism through recombinant DNA techniques.

Several products derived from recombinant DNA technology are commercially available worldwide. GMO products already on the market include human insulin, somatropin and transgenic varieties of crops, such as maize, soybeans, cotton and common beans. The United States, Brazil and Argentina are among the principal countries engaged in the commercial production and marketing of GMOs (James, 2010).

The emergence of genetic engineering in the early 1970s in California, USA, with the isolation, introduction, and expression of the insulin gene in *Escherichia coli* provoked a strong reaction from the scientific community all over the world, which led to the Asilomar Conference in 1974. At that time, the scientific community proposed a moratorium on genetic engineering. They argued that rules and safeguards should be established to ensure the use of genetic engineering techniques without risking human life and the environment. In a relatively short period of time, biosafety regulations were developed for the appropriate use of these technologies in the laboratory. After over 35 years of research on and commercial use of biotechnology, there have yet to be any reports about the adverse effects of the use of genetic engineering on human and animal health or the environment. Therefore, to ensure the appropriate generation and utilization of this technology, biosafety regulations and monitoring mechanisms have been developed in different countries around the world. Several field tests with transgenic varieties have been performed in the USA, Argentina, Bolivia and Chile since 1991. However, in Brazil, these tests only began in 1997.

2. Transgenic plants and their advantages

In the future, there will be difficulties in meeting the food demand in developing countries due the increasing trends of food prices and population growth. Therefore, it is necessary to employ new technologies, such as the use of transgenic varieties, to increase the productivity per unit area, especially in the developing nations. Qaim and Zilberman (2003) reported that in developing nations (i.e., China, India and Sub-Saharan Africa) farmers can achieve a greater than 60% grain yield advantage by using transgenic plants modified with the *Bt* gene instead of conventional varieties. In their work, these authors showed that the yield advantage comes solely from the impact of the *Bt* gene on the control of insect pests.

Biotechnology research currently plays a key role in food production because it helps to increase productivity, improve the nutritional quality of agricultural products and reduce production costs. Qaim and Zilberman (2003) reported that using the *Bt* gene for the control of insect pests gave a US $ 30 per ha advantage over conventional cotton. It is also advantageous because it allows reduction of the use of highly hazardous chemicals, such as organophosphates, carbamates, and synthetic pyrethroids, which belong to international toxicity classes I and II.

The commercialization of transgenic plants in Brazil has also strongly affected the agrochemical sector, which has annual profits of approximately 20 billion US dollars. Of this, approximately 8 billion US dollars per year corresponds to pesticides used for the control of diseases, insects and weeds. In some cases, the cost of pesticides in relation to the total cost of production reaches approximately 40%, as in cotton. However, varieties developed by genetic engineering that are tolerant to herbicides and resistant to insects, fungi, bacteria and viruses have led to reductions in the cost of agricultural production and, consequently, reduction of the impact of agrochemical wastes that have an adverse effect on the environment and human health.

The findings described above, obtained in developed and developing nations, demonstrate the contribution of transgenic plants to increasing the productivity per unit area to fulfill the increasing demand for agricultural products to feed the growing population of the world.

GMO technology is currently widely employed throughout the world. The global area of biotechnology crop coverage in 2010 reached approximately 148 million ha in 29 countries on five continents. The major biotechnology crops cultivated worldwide are maize, soybean, canola, cotton, sugar beet, alfalfa and papaya (James, 2010). The geographic distribution of biotech crops throughout the world is presented in Table 1.

3. Food biosafety

The food safety of transgenic plants is assessed in accordance with risk analysis. This methodology was initially developed with the aim of assessing deleterious effects on human health arising from potentially toxic chemicals present in food, pesticide residues, contaminants and food additives and was subsequently applied in assessing the food safety of GM plants.

One of the main foundations of risk analysis methodology is that transgenic plants are not inherently more dangerous than conventional crops; i.e., the potential health risks that may be associated with a transgenic variety are not because it is GM but rather are related to the possible chemical changes that may result from genetic modification (Konig et al., 2004). For example, a genetically modified common bean expressing an allergenic protein from the allergen Brazil Nut (*Bertholletia excelsa*) was not prohibited from being produced because it was obtained by genetic engineering, but because the genetic modification was incorporated in a gene that promotes the synthesis of an allergenic protein in this variety.

Order	Country	Area (millions of hectares)
1st	USA	66.8
2nd	Brazil	25.4
3rd	Argentina	22.9
4th	India	9.4
5th	Canada	8.8
6th	China	3.5
7th	Paraguay	2.6
8th	Pakistan	2.4
9th	South Africa	2.2
10th	Uruguay	1.1

Table 1. The ten major producers of transgenic crops in the world (adopted from James, 2010)

In general, most transgenic plants are modified to synthesize proteins that are absent in conventional varieties. These proteins are encoded by a transgene and introduced precisely for the purpose of conferring the desired trait. However, beyond this difference, other biochemical changes may result from the introduction of a transgene, and all of this is investigated during risk analysis.

In the case of transgenic plants, risk analysis is performed by comparing them with their non-GM counterparts, which are considered to be safe on the basis of their usage records. In risk analysis, instead of attempting to identify every hazard associated with the GM variety, one can seek to identify only new hazards that are not present in the traditional variety.

This type of comparative study is referred to as substantial equivalence analysis and is based on comparison of the biochemical profile of the transgenic variety with the conventional variety. The GM variety can be classified as substantially equivalent or substantially non-equivalent.

At this point, it should be noted that food security assessment of a GM plant is not restricted to applying the concept of substantial equivalence. This constitutes only the starting point for this assessment, and it aims to identify differences that will be analyzed later. Further analyses include allergenicity and toxicity tests performed *in silico*, *in vitro* and *in vivo* in animal models (i.e., rodents, birds, fish, and other species) to assess toxicity levels. In these tests, the LD_{50} (lethal dose in 50% of cases) is generally determined as an indicator of acute (i.e., short-term) toxicity.

The risk assessments are performed in three steps (Borém and Gomes, 2009):

Step 1. Risk assessment: This step can be defined as the evaluation of the probability of adverse health effects arising from human or animal exposure to a hazard. Risk assessment consists of four segments:

i.   Hazard identification, which entails the identification of biological, chemical and physical hazards found in food that may cause adverse health effects;

ii.   Hazard characterization, which entails an evaluation of an identified hazard in qualitative and quantitative terms and often involves the establishment of a dose-response relationship due to the magnitude of exposure (dose) to a physical, chemical, or biological hazard and the severity of adverse health effects;

iii.   Exposure assessment, which entails a quantitative and qualitative assessment of the likelihood of ingestion of physical, chemical and biological agents through food;

iv.   Risk characterization, which entails a qualitative and quantitative estimation of the likelihood and severity of an adverse effect on health based on identification and hazard characterization and on exposure assessment.

Step 2. Risk management: Risk management is measured from the results of risk assessment and other legitimate factors to reduce risks to the health of consumers. Measure of risk management may include labeling, imposition of conditions for marketing approval and post-trade monitoring.

Step 3. Risk communication: Risk communication includes the information exchange that must occur between all stakeholders, including the government, industry, the scientific community, media and consumers. It should occur throughout the assessment and risk management processes and should include an explanation to the public of the decisions made, ensuring access to documents obtained from the risk assessment and, at the same time, respecting the right to safeguard the confidentiality of industrial information.

Food biosafety analyses performed by different national and international organizations, such as the World Health Organization, the International Council for Science, the United Nations Food and Agriculture Organization, the Royal Society of London and the National Academies of Sciences from Brazil, Mexico, India, the United States, Australia, and Italy have demonstrated that transgenic varieties can be considered safe for human consumption.

4. Environmental biosafety

Similar to food risk assessment, environmental risk assessment considers three important points: the possibility, probability and consequences of a hazard, which should always be assessed on a case-by-case basis. This means that, following the identification of a possible danger, you should consider whether that danger is possible, if it is likely and, if it were to occur, what the result would be (Conner et al., 2003).

In the specific case of risk assessment for GM plants, a fourth point should also be considered: the risks of non-adoption of this technology.

An essential element in any risk assessment is the establishment of correct benchmarks. As described for the assessment of food security, a GM crop plant is compared with its non-GM counterpart. Similarly, the environmental impact of transgenic plants should be evaluated in relation to the impact caused by conventional varieties.

These principles are essential for providing guidance regarding which tests should be conducted and what questions should be answered to generate information that will assist in making the decision to use or not use a specific transgenic variety. Failure to follow these principles can result in unnecessary and unhelpful evaluations in risk assessments.

For example, the cultivation of insect-resistant transgenic cotton in Brazil has raised concerns about gene escape, i.e., the possibility of the transgenic variety crossing with wild species of the genus *Gossypium* that are native in Brazil and thus sexually compatible with cultivated cotton (Freire and Brandão, 2006). The main issue is the possibility of the pollen of transgenic cotton plants fertilizing wild cotton. The offspring of such crosses could have consequences for the maintenance of genetic diversity, although this point remains very controversial, as several research groups do not believe that the introduced gene would produce any adaptive advantage when exposed to the natural environment.

Gene escape from transgenic plants can occur in three main ways:

i. When the transgenic plant becomes a weed or an invasive species (e.g., for crops with weed-like characteristics, such as sunflower, canola, and rice), the transgenic gene found in the transgenic plant may allow the crop to become weedier and more invasive;
ii. Intraspecific and interspecific hybridization, such as when transgenic DNA is transferred by crossing to other varieties of cultivated species and wild species, respectively;
iii. When transgenic DNA is asexually transmitted to other species and organisms.

For a gene to escape and be transferred to different species, certain conditions are necessary:

i.   The two parental individuals must be sexually compatible;
ii. They must be located in neighboring areas and with flowering overlap between the two parental types;
iii. A sufficient amount of viable pollen must be present and transferred between individuals;
iv. The resulting progeny should be fertile and ecologically adapted to environmental conditions where the parents are located.

To avert gene escape from transgenic varieties to conventional varieties, isolation distance should be maintained. For example, maize is a wind-pollinated species, and the distances that pollen can travel depend on the wind pattern, humidity and temperature. In general, fields with transgenic varieties should be isolated from other conventional varieties with a distance of at least 200 m (Weeks et al., 2007). The risk of gene escape from soybeans and maize to wild relatives in Brazil is considered by most scientists to be small or nonexistent.

The risk of gene escape associated with transgenic soybeans in China and maize in Mexico to their wild relatives are different because China and Mexico are the centers of diversity of the respective species.

Additionally, transgenic crops may have an effect on a non-target organism. Evidence has shown that the lethal dose (LD_{50}) of *Bt* varieties for beneficial insects, such as bees and ladybugs, is far higher when such insects are exposed to fields of these transgenic varieties. Some studies have also reported the safety levels of *Bt* varieties for the monarch butterfly (Tabashnik, 1994; Tang, 1996).

A study under the auspices of the European Union addressing the environmental impacts caused by cultivation of GM crops was conducted for 15 years (1985-2000) involving 400 public research institutions and reached the following conclusion: "Our research shows that, according to standard risk assessments, GM organisms and their products do not present risks to human health or the environment. In fact, the use of more precise technology and conducting the most accurate analysis possible during the regulatory process associated with these varieties make these products even more secure than their conventional forms (European Union, 1999 – available at http://bio4eu.jrc.ec.europa.eu/documents/FINALGMcropsintheEUBIO4EU.pdf).

5. Biosafety regulations

The need for official regulation of genetically modified organisms became more evident in the mid-1980s, when biotechnology companies sought permission to perform research on genetically modified organisms.

Currently, implementation of biosafety rules is determined on a case-by-case basis around the world based on technical and scientific data, with transparent decision-making and consistency, building public confidence. There is no international standard established, and each country is responsible for creating its own regulations for research, trade, production, transport, storage and disposal.

5.1 Regulatory agencies

Field testing of the first transgenic plants began in the early 1980s. To date, there have been more than 25,000 field tests performed worldwide, half of which have been in the United States and Canada. In South America, the greatest number of releases occurred in Argentina. The commercialization of GM crops began in 1994, with tomatoes genetically engineered by Calgene. Transgenic varieties of soybean, maize, cotton, canola and papaya, among other crops, already represent a significant share of agriculture in the United States, Brazil, Canada, and Argentina. These varieties have been modified for resistance to insects and viruses and tolerance to herbicides. Field testing and laboratory evaluation of GMOs have been performed by regulatory agencies in each country to evaluate the risks of the GMOs to human and animal health and the environment.

In the United States, the agencies that examine the safety of genetically modified varieties include the Environmental Protection Agency (EPA), the Food and Drug Administration (FDA), and the Animal and Plant Health Inspection Service (APHIS) of the United States Department of Agriculture (USDA).

The USDA-APHIS regulates field tests of both plants and genetically modified microorganisms. This is the agency that reviews the licensing procedures for field testing by industry, universities and nongovernmental organizations (NGOs). The processes related to the agricultural and environmental safety of herbicide organisms, such as Roundup Ready™ (RR) soybeans, are also reviewed by USDA-APHIS.

The FDA evaluates the safety and nutritional aspects of genetically modified varieties that are used for human food and feed for animals. The FDA guidelines are based on the fact that food derived from GMOs must meet the same rigorous safety standards required for conventional foods.

The EPA is responsible for ensuring the safety of GMOs and varieties that produce pesticide elements and chemical and biological substances for distribution, consumption and trade. Under U.S. law, the jurisdiction of the EPA is limited to pesticides. For example, a plant that has been genetically modified to resist insects falls within its jurisdiction, but not a plant modified to resist drought. Plant resistance to a pest is under the authority of the EPA because the plant produces a substance that acts as a pesticide. In contrast, drought resistance may be due to factors such as deeper roots, and this transgenic plant would be subject to regulation by the USDA-APHIS.

With respect to pest-resistant varieties, the EPA has four categories of analysis: product characterization, toxicology, effects on non-target organisms and disposal in the environment. The characterization of a product includes a review of its origin and how the transgene is expressed in living organisms, the nature of the pesticide, the modifications introduced to the trait (compared with what is found in nature) and the biology of the receiving plant. To analyze the toxicology, the level of acute oral toxicity of the pesticide substance is evaluated in rats. For proteins toxic to insects, the EPA also requires a digestibility test, which evaluates the time required for the protein to be digested by gastric and intestinal juices. The EPA also analyzes the allergenicity of the protein. With regard to environmental impacts, the agency examines the exposure and toxicity of the transgenic plant to non-target insects and beneficial insects.

The regulation of biosafety is governed through local agencies in each country, such as Health Canada and the Canadian Food Inspection Agency in Canada; the Ministry of Agriculture, Forestry and Fisheries and the Ministry of Health and Labour in Japan; and Conabia in Argentina.

Several countries in Latin America, including Brazil, Argentina, Chile, Mexico and Venezuela, have established biosafety rules through specific legislation to regulate the use of genetic engineering and the release of the products of this technology into the environment. In Brazil, these rules are guided by Federal Law 11,105, enacted on March 24, 2005. This law also created the National Board of GM Biosafety (CTNBio), the National Biosafety Council (CNS), also known as the Council of Ministers, and the Biosafety Information Service (SIB). Fontes (2003) discussed the regulatory and legal concerns in Brazil in detail.

CTNBio is composed of 27 members, and their backups appointed by the scientific community, have deep scientific knowledge in the areas of biotechnology associated with humans, animals, plants and the environment, along with representatives of the several Ministries. CTNBio has developed a set of Normative Resolutions that now regulate most aspects of modern biotechnology in the country. To date, the Board has authorized the commercial release of 31 GM events for use in agriculture and vaccines for husbandry in Brazilian territory.

CTNBio analyzes the requests that are forwarded to it by issuing opinions that are specific to each transgenic target of evaluation. Before any GMO product is released for planting, trade or use, it must have been subjected to analysis of possible risks to humans, animals and the environment. The results of these tests are evaluated by CTNBio, which then makes a recommendation for release of GMOs that do not pose a risk to human or animal health or the environment. Genetically modified products suspected to have any adverse effect on human or animal health or the environment are banned from commercial use by CTNBio.

5.2 Other recently developed techniques to avoid environmental and human health concerns

5.2.1 Chloroplast engineering

The chloroplast is the principal organelle of plant cells and eukaryotic algae where photosynthesis is carried out. This is the site where food production starts. Therefore, incorporation of genes in the chloroplast genome and their expression provide a relative advantage compared with expression in the nucleus (Wang et al. 2009). The existence of many copies of chloroplast plastids, in some cases up to 10 thousand per cell, mean that chloroplast DNA comprises approximately 10-20% of cellular DNA. The other advantage of chloroplast engineering is that it is associated with greater visibility of the effects of the transgene on the production of proteins and carbohydrates. In addition, a chloroplast transgene has an advantage over a nuclear DNA transgene because it reduces the impact of contamination of the transgene via pollen to wild relatives and other similar species. This reduces the risk of developing weeds resistant to toxins and herbicides. Furthermore, the expression of a transgene in chloroplast-transgenic plants is more stable than a nuclear transformant because transgenes integrate into the chloroplast genome by homologous recombination (Elizabeth, 2005). The other advantage of chloroplast transformation is the possibility of transforming several transgenes under the control of one promoter. Chloroplast transformation has been used by different researchers to develop cultivars resistant to diseases and pests, weeds, and abiotic stresses (Wang et al., 2009, Wani et al., 2010).

Chloroplast engineering has several applications in the fields of medicine, biology and agriculture. This technique is used for improvement of plant traits, such as resistance to biotic and abiotic stress (Rhodes and Hanson, 1993), introduction of insect-resistant transgenes into crop plants (Dufourmantel et al., 2005), biopharmaceutical production (Daniell et al., 2001), metabolic pathway engineering (Lossl et al., 2005), and research on RNA editing (Hayes et al., 2006). A detailed review on chloroplast engineering was presented by Wang et al. (2009).

5.2.2 Producing marker-free transgenic plants

The objective of developing marker-free transgenic plants is of great urgency, as most of the transgenic plants developed contain the marker gene used during the selection phase. Selection genes include resistance to ampicillin (or other antimicrobials) and herbicides. The existence of this type of gene in the environment has raised a great deal of concern from environmentalists and consumer protection groups, as this might have unpredictable consequences for human and animal health. In specific cases, an herbicide-resistant gene used in the selection process could pass to weeds, resulting in the development of weeds that are resistant to herbicides. With respect to consumers, if a gene resistant to antibiotics is present in food products, it could spread to the human population, though there is no evidence for this at the moment. Therefore, developing appropriate methods of selection for transgenes without transferring the selectable marker to the environment will increase the commercialization of transgenic plants in the world by reducing the cost of developing and commercializing genetically modified crops and will make consumers more comfortable.

The currently available transformation techniques are not efficient in transforming a number of genes responsible for quantitative traits and other important agronomic traits in a single transformation. This makes joint gene transformation in a target organism impossible and increases the cost of the transformation and development of transgenic organisms. According to Puchta (2003), there are four possible ways to avoid marker genes from genetically modified crops: avoiding the use of selectable marker genes; employing marker genes with no harmful effects; joint transformation of the target trait gene and marker gene, followed by their segregation; and removing the selectable marker gene from the gene of interest (the transgene) through successful site-specific recombination or homologous recombination. Among the above four techniques used to eliminate marker genes from transgenic plants, the fourth one has recently received more attention because of its efficiency and acceptance.

The specific-site recombination system involves the Cre protein and two lox sites within the transgene construct. This system is used in a number of different types of genome manipulations. Two lox sites in direct orientation are required for the excision of the marker gene (Russell et al., 1992), and excision is performed with the expression of Cre. Elimination of the marker gene from the target gene in the transgenic plant occurs through site-specific recombination, in which the two lox sites are required for removing the marker gene from the plant genome by the use of the Cre recombinase. This technique has two principal advantages: it requires only a single round of genome manipulation; and it reduces the time required to obtain a marker-free transgenic plant (Thomason et al., 2001).

6. Detection of genetically modified organisms

There are several reasons to support detailed research on transgenic organisms, especially to allow their easy and quick identification among conventional organisms. The ability to identify a GMO is strongly related to animal, human and environmental biosafety, and it is within the rights of the consumer to know what he/she is eating. In the case of Brazil, which is the second largest transgenic producer, local legislation specifies that all food containing more than 1% material derived from GMOs should have a label indicating the presence of a GMO in its composition. The presence of genetically modified seeds in conventional seed samples has become a growing problem for international trade and may result in severe consequences for food exporters, such as Brazil. For compliance with laws and GMO regulatory measures to be effective, it is necessary to apply techniques that enable the sensitive, reliable detection and quantification of GMOs. The following section will describe the techniques used for the detection of GMOs.

6.1 Techniques for GMO detection

The genetic modifications introduced into an organism should be well known to better understand what techniques can be employed in the detection of GMOs. The basic structure of an exogenous DNA sequence inserted into a GMO is composed of three main elements, as described by Conceição and co-workers (2004): the promoter region, which is responsible for gene transcription; the gene itself, which defines the desired characteristic; and the terminator region, which is responsible for transcription termination. All detection systems are based on the elements present in the DNA sequence inserted into the GMO, either through direct detection of an exogenous DNA molecule inserted in the genome or indirectly through the protein product and by-products resulting from expression of the DNA insert.

6.1.1 Direct assays: Detection of the presence of exogenous DNA

6.1.1.1 PCR

The polymerase chain reaction (PCR) was developed by Mullis and Faloona in 1987, and it is the main technique used in molecular biology laboratories to detect GMOs. The technique is based on the replication of specific sequences of exogenous DNA. For this purpose, small pieces of DNA known as primers bind to exogenous DNA present in the GMO, and during the PCR amplification process, which is catalyzed by the DNA polymerase enzyme, thousands copies of the specific sequence are produced. Copies of DNA produced by PCR are easily visualized by electrophoresis in agarose gels. A DNA intercalating agent (e.g., ethidium bromide) is used for visualization of DNA bands present in the gel. If the same primers are used in a non-transgenic organism, the band will not displayed in an agarose gel, as there will be no detection of copies of exogenous DNA because it is not present in the wild or conventional organism.

The PCR technique is very specific, sensitive and safe and is able to detect both events of genetic modification (Bertheau et al., 2002; Giovannini and Concillo, 2002) and distinguish events associated with different gene constructs expressing the same protein (Yamaguchi et al., 2003). However, this technique also presents some limitations, such as: 1) the difficulty involved in designing primers, as it is necessary to know the genetic sequence of the DNA introduced into the GMO, and this information is usually confidential (Holst-Jensen et al., 2003); 2) the need for appropriate equipment and trained personnel; 3) the relatively high cost because the test is specific for each genetic alteration introduced; and 4) the special care required to avoid sample contamination (Miraglia et al., 2004, Yamaguchi et al., 2003).

The quality of DNA extracted is crucial to the success of the PCR method. There are several procedures described in the literature for DNA extraction from leaves, seeds, and even processed foods. The CTAB (cetyltrimethylammonium bromide) method is widely used in molecular biology laboratories. There are also several commercial kits that employ silica resin with high affinity for DNA molecules. Poor quality, degraded or low purity DNA can negatively influence the success of PCR, preventing the identification of foreign DNA in the sample. This low purity DNA can occur from the presence of inhibitors in DNA extracts (e.g., proteins, polysaccharides, and polyphenols) that hinder the annealing of primers to target DNA and/or inhibit the activity of the DNA polymerase enzyme (Ahmed, 2002). Consequently, the reaction will occur with a low efficiency or may not occur at all. For quality control in PCR, it is always necessary to use a standard reaction (control) that evaluates the quality of the extracted DNA and avoids false-negative results. The default reaction may be performed using specific primers for any known endogenous gene.

The limit of detection for PCR is between 20 picograms and 10 nanograms of exogenous DNA. Thus, it is possible to detect a single genetically modified seed among 1,000 to 10,000 conventional seeds (Luthy, 1999).

6.1.1.2 Real-time PCR

Various PCR techniques are used in molecular analysis of transgenic events for different purposes. The real-time qPCR (quantitative PCR) technique was developed and has been used in the identification of GMO events and products. Real-time qPCR is used not only for the detection of specific DNA but also for the quantification of copy number of a particular target DNA sequence inserted in a GMO. The difference between conventional PCR and qPCR is the specificity and sensitivity of the latter method. In real-time qPCR, equipment capable of detecting the fluorescence emitted by the reaction during each cycle of amplification of the target DNA molecule is used whereas in conventional PCR, the result is visualized in an agarose gel after 30-45 cycles of amplification. Real-time qPCR is monitored from the first cycle of amplification until the last one by detecting the fluorescence emitted. Then, it is possible to recognize the exact time (i.e., cycle) at which the amplification of the target molecule can be detected (Figure 1). These data allow inference of the number of copies of the transgene present in the GMO based on an endogenous control reaction. Basically, when the number of amplification cycles required to detect the emitted fluorescence is reduced, the copy number of the transgene inserted in the GMO is higher (an inverse correlation). The sensitivity of the method is based not only on the uptake of the fluorescent signal but also on how fluorescence is emitted during the reaction. Most real-time PCR applications require only one fluorescent agent for double-stranded DNA. However, some applications require greater specificity, such as the TaqMan® system.

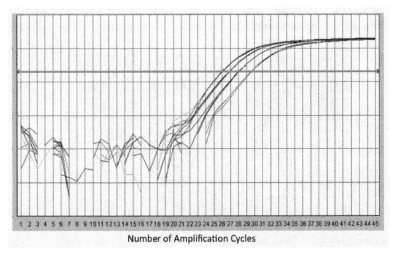

Number of Amplification Cycles

Fig. 1. Real-time qPCR. The curves in the graph show the progress of target molecule amplification. Capture of the fluorescent signal occurs throughout all PCR cycles. Source: Camargo (2009).

Real-time qPCR using the TaqMan® detection system employs a pair of primers and a probe labeled with a fluorophore (a marker that emits fluorescence when stimulated). These three oligonucleotides (two primers and a probe) are specific to the target sequence, thereby contributing to the greater specificity of the technique. In the process of primer amplification; the probe binds to the target DNA molecule, and a DNA polymerase enzyme initiates polymerization of the new molecule. When the enzyme meets the probe linked to the target DNA, it severs the probe and allows fluorescence to be emitted and then detected by the equipment. Therefore, the process of fluorescence emission is dependent on the joint action of four elements: two primers, one probe and the DNA polymerase enzyme.

There are factors related to amplification conditions that may adversely affect the reliability of the results obtained, such as the use of inappropriate temperatures for primer and probe annealing; low specificity of the primers and probe with respect to annealing to the DNA template; and inappropriate conditions for the activity of the polymerase enzyme (e.g., unadjusted salt concentration or pH of the buffer).

6.1.1.3 Southern blotting

The Southern blot technique, described by Southern in 1975, is also frequently used in laboratories to detect specific fragments of exogenous DNA integrated into the genomic DNA of a transgenic organism and its products. This technique essentially consists of five steps: 1) extraction and digestion of genomic DNA with one or more restriction enzymes; 2) separation of DNA fragments by electrophoresis in an agarose gel; 3) transfer and fixation of DNA present in the gel to a nitrocellulose or nylon membrane; 4) hybridization of DNA present in the membrane against a DNA probe that has sequence homology to the target DNA; and 5) visualization by autoradiography or colorimetry.

Southern blot analysis is very reliable and is considered molecular evidence of the integration of exogenous elements in a GMO genome. In addition, it is also possible to estimate the number of copies that were introduced into the genome of the recipient organism with this method (Figure 2).

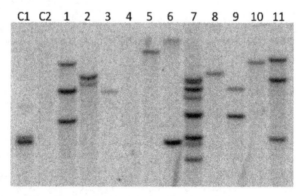

Fig. 2. Southern blot analysis. Analysis of transgenic events using the Southern blot technique. C1) Positive control (transgenic event); C2) negative control (wild plant); 1 to 11) transgenic events in the analysis. In this case, the number of bands refers to the copy number of target DNA sequences present in the genome of each event analyzed. Source: Camargo (2009).

However, this technique is associated with the limitation of requiring a large amount of genomic DNA (between 20 to 40 micrograms), which is sometimes difficult to obtain depending on the type of sample used. Other limitations to this technique are its high cost, operational complexity, long period required to perform the experiment and obtain results, the use of radioactive probes, and that it requires an appropriate infrastructure and adequate training for the handling and storage of radioactive products and waste disposal.

6.1.2 Indirect assays: Detection based on the presence of RNA

The genetic information in DNA must be translated into protein to be effective and have effects in an organism. The translation of information from DNA to protein occurs only because of the previous transcription of DNA into molecules of messenger RNA (mRNA). mRNA synthesis can be considered as the intermediate stage of the process of transferring the information contained in DNA and reflects the level of transcription activity, as the presence of mRNA is directly related to gene expression.

There are different molecular techniques that can be used in studies of gene expression. These include the northern blot and RT-PCR (reverse transcription - polymerase chain reaction) techniques. These techniques can be employed to determine the gene expression in different tissues and/or different stages of development of the organism under study and to monitor the gene expression of exogenous DNA in GMOs.

6.1.2.1 Northern blotting

The northern blot technique, also referred to as an RNA blot, was developed to study gene expression through the detection of RNA molecules present in a sample (Alwine et al., 1977). The execution of northern blotting is very similar to the Southern blot technique and basically consists of 5 steps: 1) total RNA extraction, 2) separation of RNA fragments by electrophoresis in an agarose gel, 3) transfer and fixation of the RNA present in the gel to a nitrocellulose or nylon membrane, 4) hybridization of the RNA present in the membrane against a DNA or RNA probe with homology to the target RNA sequence, and 5) revelation by autoradiography.

One of the most important steps in this technique is the extraction of total RNA from the sample because to obtain reliable results, it is necessary to obtain intact and pure RNA. Exercising care during the technique is much more critical than in Southern blotting because RNA degrades easily. To prevent its degradation, it is necessary to treat all objects with specific solutions to eliminate or minimize the presence of RNases, which are enzymes that are very effective in degrading RNA molecules.

6.1.2.2 RT-PCR

RT-PCR is widely used to verify gene expression by detecting mRNA molecules. This technique is based on reverse transcription of mRNA followed by PCR amplification. The reverse transcription reaction is based on the synthesis of complementary DNA (cDNA) from an mRNA molecule template by the reverse transcriptase enzyme. The product of this amplification is visualized in an agarose gel. The intensity of the bands visualized in the gel provides some indication of the amount of target mRNA present in the sample.

Quantitative RT-PCR (qRT-PCR) is a modern method based on the principles of RT-PCR (i.e., cDNA production followed by PCR). Therefore, qRT-PCR is more robust, specific and sensitive; consequently, it provides better quantitative results. The amplification progress of the target molecule is displayed in real time by capturing a fluorescent signal in more sophisticated thermal cyclers, such as those described for the qPCR technique.

6.1.3 Indirect assays: Detection based on the presence of protein

6.1.3.1 Bioassays

In most of the GM varieties commercialized to date, genes have been introduced conferring tolerance to herbicides and/or resistance to viruses, fungi or insects (Borém and Almeida, 2011). The bioassay technique for the detection of GMOs in these cases is relatively simple, inexpensive and easy to establish in both laboratories and greenhouses, but a relatively long time is required to obtain results (the results are usually obtained after a week).

Bioassays for herbicide tolerance can be conducted using a plant or even seeds. In case of plants, a dose of the herbicide is sprayed on the leaves. Then, the plants are monitored daily to verify the presence or absence of any phenotype (i.e., symptoms) resulting from the application of the herbicide (Figure 3). The leaves of plants with no tolerance to the herbicide initially become yellow and then dry (i.e., the tissue undergoes necrosis). The leaves of herbicide-tolerant plants exhibit no or few symptoms of necrosis, and the plants continue to look as healthy as before the herbicide application. In the case of bioassays performed with seeds, the seeds are germinated in a medium containing a diluted solution of the herbicide. If the seeds are tolerant to the herbicide, germination will occur, and the plant will develop normally, as is seen for the seeds of transgenic Liberty Link™ corn and Roundup Ready™ soybeans, which are tolerant to glyphosate. If the seeds are sensitive to the herbicide, no germination will be observed. Currently, bioassays for herbicide tolerance are commonly used by companies that export seeds and grains to prove the authenticity and quality of their products (Torres et al., 2003).

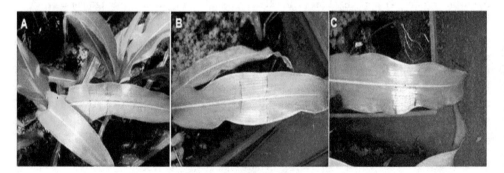

Fig. 3. Bioassay for herbicide tolerance in genetically modified maize plants overexpressing the *bar* gene. This gene encodes the enzyme phosphinothricin-N-acetyltransferase (PAT), which confers tolerance to the herbicide ammonium gluphosinate (PPT). a) Leaf of a genetically modified maize plant showing PPT tolerance after 10 days of herbicide application; b) leaf of a genetically modified maize plant showing mild PPT susceptibility after 10 days of application; c) leaf of an unmodified maize plant (negative control) showing an intense PPT susceptibility after 10 days of application. Source: Camargo (2009).

Insect-resistant plants can also be analyzed by bioassays. In this case, the bioassay can be performed by placing insects or their larval form, according to the cycle of the insect that attacks the plant, on the plants to be analyzed or even on leaf discs from the plants. Two types of information can be obtained from these experiments: 1) the mortality rate of the pest; and 2) the damage caused by the pests to the analyzed tissue. Analysis of these data will indicate whether or not the plant is resistant to the insect under investigation.

Such bioassays are based on the gene expression and phenotype of interest presented by the GMO. However, a disadvantage of this method is that the results of bioassays are generally not sufficiently definitive proof of the integration of exogenous DNA into the genome of a transgenic organism, and other evidence (i.e., results from other techniques) is necessary for its verification.

6.1.3.2 Immunoassays

Immunoassays are ideal methods for the qualitative and/or quantitative detection of specific proteins produced from an exogenous DNA sequence introduced into a GMO. The main immunoassays used for the detection and quantification of target proteins present in GMOs are the enzyme-linked immunosorbent assay (ELISA), western blotting and the lateral flow immunoassay (LFI).

6.1.3.2.1 ELISA

An ELISA identifies a target protein present in a protein extract containing a population of other proteins by using specific antibodies that bind to the target protein. The antigen-antibody reaction identifies the presence of exogenous protein in a qualitative and even quantitative form.

There are several types of ELISA, including direct ELISA, indirect ELISA, sandwich ELISA and competitive ELISA. The sandwich ELISA method, which uses two specific antibodies (Abs) for a target protein, is the most sensitive of these techniques and is used for the detection of GMOs (Yates, 1999). In this type of immunoassay, two types of antibodies specific for a target protein are used: a primary (i.e., capture) Ab, which is used to sensitize a plate to capture the target protein (i.e., the antigen, Ag) present in the sample; and a secondary (i.e., combined) Ab, usually in conjunction with an enzyme (e.g., peroxidase or alkaline phosphatase) that acts on a given substrate to produce color (in a colorimetric assay) or fluorescence (in a fluorimetric assay). The intensity of the color produced is directly related to the amount of antigen present in the sample, as the color will only occur when the target protein binds to the capture antibody on the plate, followed by binding of the enzyme-conjugated antibody to the immobilized target protein (Figure 4). Free antibodies and protein that did not form Ab-Ag-Ab complexes are discarded during microplate washing steps. Thus, there is little possibility for false-positive results to occur. For quantitative assays, a standard curve of protein at known concentrations is used.

Direct and indirect ELISA assays are similar to the sandwich ELISA, but with some modified steps. A direct ELISA assay uses only a specific antibody conjugated to an enzyme (i.e., the "secondary Ab"). Protein extract is added to a plate followed by the conjugated antibody. In the next step, the appropriate substrate is added; the enzyme then acts on it, and the reaction is revealed. An indirect ELISA assay uses a primary Ab specific for a target protein and an enzyme-conjugated Ab that is specific to the primary Ab (anti-IgG of the organism in which the primary Ab was produced, usually rabbit or mouse). In this case, protein extract is added to a plate, followed by the primary Ab. The secondary Ab is added to the reaction, recognizing the primary Ab, and then binding to the Ab-Ag complex. The development stage occurs in the same way as in the direct ELISA. In many cases, the sandwich ELISA assay is two to five times more sensitive than the direct or indirect ELISA.

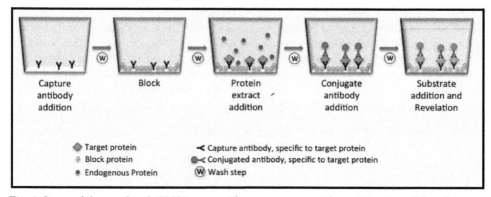

Fig. 4. Steps of the sandwich ELISA assay. The assay starts with sensitization of the plate with the capture antibody. After washing, the plate is blocked. Then, the protein extract is added. The plate is washed again, and conjugated antibody is added. After incubation, the plate is washed and then the protein present is revealed by adding the substrate. Source: Camargo (2009).

Another kind of ELISA is the competitive assay, in which the target protein in the sample and a standard protein conjugated to an enzyme compete for the binding of a capture antibody. In this type of test, the amount of target protein is inversely proportional to the colorimetric intensity produced. The more standard protein binds to the antibody, the more intense the color produced by the reaction. Some drawbacks of competitive ELISA are its restriction for use with only one specific antibody each time it is used and that it is less sensitive than the sandwich method.

The type of antibody used in an immunoassay can also contribute to the greater sensitivity and specificity of the assay. Monoclonal antibodies contribute to increasing the specificity of this type of technique while polyclonal antibodies increase sensitivity because they can recognize different epitopes of a target protein (Ahmed, 2002).

ELISA is a very sensitive, specific, robust, safe and rapid technique for the detection of GMOs in the laboratory. Moreover, it is the ideal technique for simultaneous analysis of a large number of samples under routine diagnosis.

6.1.3.2.2 Western blotting

The western blot technique is based on the separation of proteins present in the protein extract from a sample in non-denaturing polyacrylamide gels and subsequent transfer to nitrocellulose or nylon membranes. Detection of the target protein is performed by means of specific antibodies that recognize epitopes of the protein of interest. Visualization of the assay occurs through a colorimetric reaction or radiographic detection. Thus, western blot analysis combines the resolution of electrophoresis with the specificity of immunological detection (Brasileiro and Carneiro, 1998).

The electrophoretic separation of proteins in the sample usually occurs under denaturing conditions. Thus, problems of solubility, aggregation and co-precipitation of the target protein with other proteins present in the sample are eliminated (Sambrook and Russel, 2001). However, antibodies against conformational epitopes of the target protein may not recognize these epitopes when denatured.

The western blot technique is semiquantitative, specific and sufficiently sensitive to detect proteins (Brett et al., 1999). The limit of detection for the target protein depends on a number of factors, including the type of membrane and the detection system used. In most cases, this limit corresponds to approximately 20 femtomoles (10-15 moles). Thus, it is possible to detect approximately 1 nanogram of a protein with a molecular weight of 50 kDa (Brasileiro and Carneiro, 1998). In seed analysis, the minimum limit of detection is 0.25% (Yates, 1999). Western blot analysis is a laborious technique, and it is capable of analyzing only a few samples simultaneously. Therefore, western blotting is rarely used in routine analysis of GMOs. This technique is usually used to confirm preliminary results generated by other detection techniques.

6.1.3.2.3 Lateral flow immunoassay

The lateral flow immunoassay (LFI) is widely used for analysis of material still in field trials and product testing because it is practical, inexpensive and fast. Its results are obtained within 5 to 15 minutes. Other advantages of this method are that it does not require special equipment and trained personnel. However, this technique is not sufficiently robust for quantification of GM material present in a sample, but it is a very sensitive technique for the qualitative detection of GMOs (Urbanek et al., 2001).

The principle of the lateral flow immunoassay is similar to the sandwich ELISA. The detection antibody is located at the end of a strip that is inserted into a sample solution. The target protein present in the sample binds to antibody present in the strip, and the Ag-Ab complex then migrates by capillary action to the other end of the strip, where there are two capture zones. In these areas, there are specific antibodies to capture the target protein or detection antibody. When the Ag-Ab complex passes through the capture zones, there is a colorimetric reaction (Figure 5). The presence of two colored bands on the strip indicates that the test is positive (i.e., the transgenic protein is present in the sample). The presence of only one band indicates that the sample is negative (i.e., that it contains no traces of the transgenic protein, but the test was performed correctly) (Conceição et al., 2004).

IFL strips are produced commercially for the detection of a wide range of proteins used in the production of GMOs. Soybeans, maize, canola, cotton and sugar beets genetically modified to contain the endotoxin Cry (Ab) from *Bacillus thuringiensis* or the CP4-EPSPS protein from *Agrobacterium tumefaciens* can be easily analyzed using this technique (Lipton et al., 2000).

In addition to their usefulness in the detection of GMOs, immunoassays are powerful tools for assessing the expression of a transgene. It is possible to identify the location of transgene expression in a plant, in which tissues the protein is present (e.g., roots, leaves, seeds) and the ratio of expression among different tissues (i.e., tissues that show more or less expression of the transgene).

6.1.4 Alternative techniques for the detection of GMOs

Because of the increasing number of GMOs and the complexity of the genetic changes that are emerging, new techniques are being developed or improved with the goals of increasing sensitivity and reliability, lowering costs and allowing simultaneous analysis. DNA microarrays, chromatography and mass spectrometry are examples of other techniques for detecting GMOs.

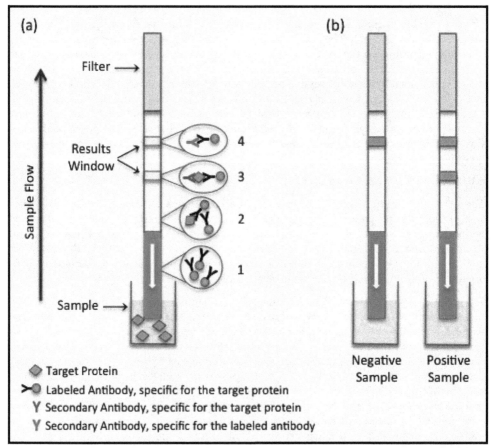

Fig. 5. Diagram of a lateral flow immunoassay (IFL). a) One end of the IFL strip is inserted into the sample. By capillarity, the sample travels up the strip toward the other end. During this run, the sample passes through a region where there are capture antibodies (1), and the target protein binds. Capture antibodies that are either complexed with the target protein or free (2) migrate and bind to specific antibodies for the target protein (3) or specific antibodies to the capture antibodies (4), respectively. b) A single band revealed in the results window indicates that the sample is negative (i.e., absence of the target protein), and two bands indicate that the sample is positive (i.e., presence of the target protein). Source: Camargo (2009).

7. Conclusions

Regulatory procedures for GM crops require extensive risk analysis on a case-by-case basis for modified organisms. Information regarding the number of copies of foreign DNA inserted into the genome, the expression level of the protein of interest, the parts of the plant in which the protein is present, the toxicity and allergenicity of the protein and its possible adverse effects for non-target organisms and for the environment is required. Thus, numerous techniques have been used to evaluate the biosafety of GMOs. The detection and identification of these organisms are also of great interest for identifying the purity of seed samples, labeling food, and trade reasons.

Knowledge regarding the genetic constitution of a GMO and the main features of each technique is essential for the implementation of these tests and to obtain accurate and reliable results with respect to detecting and evaluating a transgenic organism. In some cases, the simple integration of exogenous DNA into the host genome does not mean that the gene is being expressed and that the target protein is being produced. Thus, the combined use of more than one detection technique may be necessary for the complete assessment of GMOs.

8. Acknowledgements

The authors would like to thank CAPES (PNPD-CAPES-FINEP supply the fellowship for the first and last authors), FAPEMIG, CNPq, FINEP and EMBRAPA for financial support and researcherships.

9. References

Ahmed, FE (2002) Detection of genetically modified organisms in foods. Trends in Biotechnology 20:215-223

Alwine, JC; Kemp, DJ; Stark, GR (1977) Method for detection of specific RNAs in agarose gels by transfer to diazobenzyloxymethyl-paper and hybridization with DNA probes. Proceedings of the National Academy of Sciences USA 74:5350-5354

Bertheau, Y; Diolez, A; Kobilinsky, A; Magin, K (2002) Detection methods and performance criteria for genetically modified organisms. Journal of AOAC International 85:801-808

Borém, A; Almeida, GD; (2011). Plantas Geneticamente Modificadas. 1. Ed. Visconde do Rio Branco, 2011. 390p.

Borém, A; Gomes, WS (2009) Biossegurança e sociedade. Informe Agropecuário 30: 7-13.

Brasileiro, ACM; Carneiro, VTC (1998) Manual de Transformação Genética de Plantas. 1.ed. Brasília, 1998. 309p. (EMBRAPA-SPI/ EMBRAPA-Cenargen).

Brett, GM.; Chambers, SJ.; Huang, L; Morgan, MRA (1999) Design and development of immunoassays for detection of proteins. Food Control 10:401-406

Camargo, S. R. (2009) Detecção de produtos e organismos transgênicos. Informe Agropecuário. 30(253): 44-52.

Conceição, FR; Moreira, AN; Binsfeld, PC (2004) Detecção de organismos geneticamente modificados em alimentos e ingredientes alimentares. Ciência Rural 36:315-324

Conner, AJ; Travis, TR.; Nap JP (2003) The release of genetically modified crops into the environment Part II. Overview of ecological risk assessment. Plant J. 33, pp. 19–46.

Daniell, H., Lee, S.B., Panchal, T., and Wiebe, P.O. (2001). Expression of the native cholera toxin B subunit gene and assembly as functional oligomers in transgenic tobacco chloroplasts. J. Mol. Biol. 311: 1001–1009.

Dufourmantel, N., Tissot, G., Goutorbe, F., Garçon, F., Muhr, C., Jansens, S., Pelissier, B., Peltier, G., and Dubald, M. (2005). Generation and analysis of soybean plastid transformants expressing Bacillus thuringiensis Cry1Ab protoxin. Plant Mol. Biol. 58: 659–668.

Elizabeth P M. (2005) Engineering the Chloroplast Genome. Resonance: 94-95

Freire, EC; Brandão, ZN (2006) Biotechnology of cotton in Brazil: legal and political impediments and their effects in the competitiveness of the Brazilian cotton. In: 65th plenary meetings ICAC, 2006, Goiania-GO. 65th plenary meetings ICAC. Goiania-Go: ICAC, 2006.

Fontes, EMG (2003) Legal and regulatory concerns about transgenic plants in Brazil. Journal of Invertebrate Pathology 83: 100–103.

Giovannini, T; Concillo, L (2002) PCR detection of genetically modified organisms. Starch 54:321-327

Hayes, M.L., Reed, M.L., Hegeman, C.E., and Hanson, M.R. (2006) Sequence elements critical for efficient RNA editing of a tobacco chloroplast transcript in vivo and in vitro. Nucl. Acids Res. 34: 3742–3754.

Holst-Jensen, A; Ronning, SB; Lovseth, A; Berdal, KG (2003) PCR technology for screening and quantification of genetically modified organisms (GMOs). Analytical and Bioanalytical Chemistry 375:985-993

James, C (2010) Global status of commercialized transgenic crops: 2010. Available at: http://www.isaaa.org/publications

Lipton, CR; Dautlick, JX; Grothaus, GD; Hunst, PL; Magin, KM; Mihaliak, CA, Rubio, FM; Stave, JW (2000) Guidelines for the validation and use of immunoassays for determining of introduced proteins in biotechnology enhanced crops and derived food ingredients. Food Agricultural and Immunology 12:153-164.

Lossl, A., Bohmert, K., Harloff, H., Eibl, C., Muhlbauer, S., and Koop, H.-U. (2005). Inducible trans-activation of plastid transgenes: Expression of the R. eutropha phb operon in transplastomic tobacco. Plant Cell Physiol. 46: 1462–1471.

Lüthy, J (1999) Detection strategies for food authenticity and genetically modified organisms in food. Food Control 10:359-361

Konig, A; Kocburn, A; Crevel, RWR; Debruyne, E; Grafstroem, R; Hammerling, U; Kimber, I; Kinudsen, I; Kuiper, HA; Peijnenburg, CM; Penninks, AH; Paulsen, M; Schauzu, M; Wall, JW (2004) Assessment of the safety of foods derived from genetically modified (GM) crops. Food and Chemical Toxicology, vol. 42, no. 7, p. 1047-1088.

Miraglia, M; Berdal, KG.; Brera, C; Corbisier, P; Holst-Jensen, A; Kok, EJ; Marvin, HJP; Schimmel, H.; Rentsch, J; Van Rie, JPPF; Zagon, J (2004) Detection and traceability of genetically modified organism in the food production chain. Food Chemical Toxicology 42:1157-1180

Mullis, KB; Faloona, FA (1987) A polymerase catalyzed chain reaction. In: WU, R (Ed.). Recombinante DNA: part F. San Diego: Academic Press 335-350 (Methods in Enzimology, 155).

Puchta H. 2003. Marker-free transgenic plants. Plant Cell, Tissue and Organ Culture 74: 123-134

Qaim, M; Zilberman, D (2003) Yield Effects of Genetically Modified Crops in Developing Countries. Science 299:900-902.

Rhodes, D., and Hanson, A.D. (1993). Quaternary ammonium and tertiary sulfonium compounds in higher plants. Annu. Rev. Plant Physiol. Plant Mol. Biol. 44: 357-384.

Russell S H, Hoopes J L, Odell J T. (1992) Directed excision of a transgene from the plant genome, Mol Gen Genet, 2: 49-59

Sambrook, J; Russell, DW (2001) Molecular Cloning: A Laboratory Manual. Cold Spring Harbor Laboratory Press, Cold Spring Harbor, New York

Southern, EM (1975) Detection of specific sequences among DNA fragments separated by gel electrophoresis. Journal of Molecular Biology 98:503-517

Tabashnik BE. Evolution of resistance to Bacillus thuringiensis. Annual Review of Entomology. 1994;39:47-79.

Tang JD, Shelton AM, Van Rie J, de Roeck S, Moar WJ, et al. (1996) Toxicity of Bacillus thuringiensis spore and crystal protein to resistant diamondback moth (Plutella xylostella). Appl. Environ. Microbiol. 62:564-69

Thomason LC, Calendar R & Ow DW (2001) Gene insertion and replacement in Schizosaccharomyces pombe mediated by the Streptomyces bacteriophage phiC31 site-specific recombination system. Mol. Genet. Genom. 265: 1031-1038

Torres, AC; Nascimento, WM; Paiva, SAV; Aragão, FAS (2003) Bioassay for detection of transgenic soybean seeds tolerant to glyphosate. Pesquisa Agropecuária Brasileira 38:1053-1057

Urbanek-Karlowska, B; Fonberg-Broczek, M; Sawilska-Rautenstrauch, D; Badowski, P; Jedra, M (2001) Usefulness of an immunoassay test trait for detection of genetically modified Roundup ready soybean in food products. Rocz Panstwowy Zaklad Higieny 52:313-320

Yamaguchi, H; Sasaki, K; Umetsu, H; Kamada, H (2003) Two detection methods of genetically modified maize and the state of its import into Japan. Food Control 14:201-206

Yates, K (1999) Detection Methods for Novel Foods Derived from Genetically Modified Organisms. Food Control 10:339-414

Wani SH, Haider N, Kumar H, Singh NB (2010) Plant Plastid Engineering. Curr Genomics 11: 500-512.

Wang HH , Yin WB, Hu ZM. 2009. Advances in chloroplast engineering. Journal of Genetics and Genomics 36: 387-398

Weeks, R; Allnutt, T; Boffey, C; Morgan, S; Bilton, M; Daniels, R; Henry, C (2007) A study of crop-to-crop gene flow using farm scale sites of fodder maize (*Zea mays* L.) in the UK. Transgenic Research 16: 203-211.

9

Transgenic Plants – Advantages Regarding Their Cultivation, Potentially Risks and Legislation Regarding GMO's

Pusta Dana Liana
University of Agricultural Sciences and Veterinary Medicine Cluj-Napoca
Romania

1. Introduction

Transgenic plants are the results of modern biotechnology. Biotechnology represents "any technological application that uses biological systems, living organisms, or derivatives thereof, to make or modify products or processes for specific use" as it was defined by the United Nations Convention on Biological Diversity. So, modern biotechnology represents the technology by which the genetic material of an organism is modified. This technique can be applied to micro-organisms, plants, animals and even to humans. This modern technology by which the genes are transferred from one organism to another is called genes technology, genetic modification, genetic engineering or bioengineering and the transferred genes are called transgenes. The receptor organism, which is transformed by the insertion of the new genes into its genome, is named transgenic organism or genetic modified organism. The donor and receptor organisms may belong to very different species which can not cross by natural means. This fact makes the differentiation of the modern biotechnology by the conventional amelioration techniques. So, the transgenic organisms can not be obtained free in nature, in natural breeding, they can be obtained just with the help of modern biotechnologies.

Genetic engineering appeared as an own scientific discipline at the beginnings of '70 and till today it has many definitions, but the most comprehensive looks to be the following: genetic engineering represents an ensemble of methods and technologies made "in vitro" with genes, chromosomes and some times with entire cells, on the purpose of "building" of some new genetic structures with premeditated hereditary properties (Popa L. and col., 1982, as cited Vlaic A., 1998).

Genetic engineering has created two new concepts such as: genetic modified organism and the term of transgenesis.

The concept of genetic modified organism represents a legislative term which groups all the organisms resulted by genetic engineering techniques, obtained by different methods of genetic recombinations.

The transgenic organism term implies the incorporation of some exterior genes, considered as useful, obtained by the recombinant DNA technology in the genome of some zygotes, resulting transgenic organisms with modified genetical properties. This gene transfer is called transgenesis because it presumes crossing the barriers (transgression) meaning the gene transfer between different species, especially belonging to different genus (a human gene transferred in the bacteria genome, a bacterial gene transferred into the genome of a plant or animal).

Practically the term of genetic modified organism designate every organism of which genetic patrimony was modified using the specific methods of genetic engineering.

2. Obtaining of the genetic modified plants (transgenic plants)

The genetic manipulation of plants has been going on since the dawn of agriculture, but until recently this has required the slow and tedious process of cross-breeding varieties. Genetic engineering promises to speed the process and broaden the scope of what can be done.

Progress is being made on several fronts to introduce new traits into plants using recombinant DNA technology. Recombinant DNA is DNA that has been created artificially. DNA from two or more sources is incorporated into a single recombinant molecule.

In order to obtain transgenic plants it is necessary to perform the following steps:

a.   transfer of the gene of interest in the host cell;
b.   selection of the host cells which have integrated the transgenes in their genome;
c.   regeneration of some whole plants, starting from plantulas obtained by „in vitro" cultivation of the host cells;
d.   cultivation of the transgenic plants in protected environments (green houses, etc.).
e.   experimental cultivation of the new plants in the fields.

Each step of the operation requires some special aspects, such as:

a.   transfer of the foreign gene of interest in the genome of the vegetal cell can be realized both indirectly or directly.

The indirect technique presume the using of a biologic vector, represented by some bacteria, plasmids or viruses, capable to introduce, naturally, a part of their DNA in the host cells, which provides new properties for the receptor plant. The direct method uses some techniques as the cellular microinjection of the recombinant DNA or the electro-perforation of the cellular membrane by electric shocks which produce micro-pores allowing the transgenic DNA to penetrate the new cell.

b.   Selection of the host cells of the gene of interest, no matter the transfer method used. The success rate is always very low and for this reason it is necessary to associate the gene of interest with a marker gene, favorising the selection of the cells in which the transgene was integrated. The most used marker genes by the transgenic seeds producers are the ones which are codifying the resistance against an antibiotic or herbicide.
c.   The regeneration and the cultivation of the transgenic plants in protected areas are made after they have been obtained *in vitro*. The plants obtained by cultivation in vitro, must be transferred into green houses or rooms with controlled climate. This process must be repeated for several generations, to control the expression of the new character, its hereditary transmission and the absence of the unwanted effects.
d.   The experimental cultivation in field has the main aim to test the behavior of the new transgenic plant in natural conditions. Another aim is to cross the transgenic plant with

the conventional elite type (with the best performances at the time), to obtain varieties of GMO with an increase productive efficiency. Theoretically speaking, at least, these experimental cultures should also evaluate the impact of the GMO regarding the environment and human health.

All the experimental cultures of transgenic plants are conditioned by an authorization issued by a group of official experts. The name of the transgenic plant, the nature of the experiment, the location and the dimensions of the cultivated experimental areas must be communicated to the local authorities and to the public. In average, about one decade passes between the first laboratory manipulations and the commercial cultivation of the plants, all these experiments being done with high costs.

Finally, we still mention that all the transgenic plants are protected by industrial type registered marks. All the farmers who buy GM seeds are obliged, by contract, not to keep a part of their crop for seeds and the penalties for nonobservance of the law are very high.

3. New properties of the transgenic plants

The obtaining and, after this, the production of the transgenic plants, with commercial aim, determined a lot of enthusiasm within the researchers. The initial enthusiasm was then followed by the opposition of the skeptics, who tried, during the time, to argue against the cultivation of transgenic plants.

The supporters of transgenic plants cultivation base their arguments on the new properties of the transgenic plants. Among these, the most important are: resistance against the herbicides (more than half of the tries), followed by the resistance against the illnesses (mostly the viral ones) and against the insects.

We will shortly describe now the properties of transgenic plants already cultivated on a large scale.

a.   Resistance against the herbicides

It is manifested by the capacity of the plant to live and to develop after it was sprayed with a strong insecticide substance. This property is due to the transfer of the *Bar* gene which determines different enzymatic actions materialized by the transformation of the herbicide into a non toxic element. In the case of transgenic plants, the resistance against the herbicides is uni-specific, so the plant is resistant only against the herbicide for which it was created and for this reason the producer of the transgenic plant delivers also the characteristic herbicide for each transgenic plant.

b.   Resistance against the insect pests

The plants having this property are permanently synthetize in their tissue an insecticide protein that determine the death of the phytophage insects. The gene codifying the resistance against the insects originates in a soil bacteria named *Bacillus thuringiensis*. *Bacillus thuringiensis* is a bacterium that is pathogenic for a number of insect pests. The lethal effect is determined by a proteic toxine which is produced. Through recombinant DNA methods, the toxin gene can be introduced directly into the genome of the plant where it is expressed and provides protection against insect pests of the plant.

c.   Resistance against illnesses

Genes that provide resistance against plant viruses have been successfully introduced into such crop plants as tobacco, tomatoes, and potatoes. By transferring the gene codifying the protein of the viral capside there were obtained plants resistant against the illnesses produced by some viruses because it blocks the propagation of the viruses in the transgenic plant.

d.   Resistance against freezing

Even not on a large scale, but there were obtained GMO resistant against freezing. This type of resistance was obtained by two methods. First, the conventional cultures (especially strawberries) were treated with transgenic "antifreezing" bacteria. The second method consisted in insertion of some genes obtained from fish living in cold water, as *Hippoglossus hippoglossus*, a fish living in the North Sea, transferred into strawberries.

e.   Improved nutritional quality

Milled rice is the staple food for a large fraction of the world's human population. Milling rice removes the husk and any beta-carotene it contained. Beta-carotene is a precursor to vitamin A, so it is not surprising that vitamin A deficiency is widespread, especially in the countries of Southeast Asia. The synthesis of beta-carotene requires a number of enzyme-catalyzed steps. In January 2000, a group of European researchers reported that they had succeeded in incorporating three transgenes into rice that enabled the plants to manufacture beta-carotene in their endosperm.

f.   Delayed maturation

This property was first conferred to tomatoes. In this case, there were also used two main methods:

-   insertion of a gene that blocks the galacturonase, an enzyme producing the fruits softening;
-   blockage of the maturating hormone synthesis; the maturation is than started by treatment with ethylene before transferring on the market.

g.   Salt tolerance

A large part of the land is so laden with salt that it cannot be used to grow most important crops. However, researchers at the University of California Davis campus have created transgenic tomatoes that grow well in saline soils. The transgene was a highly-expressed sodium/proton antiport pump that sequestered excess sodium in the vacuole of leaf cells.

Due to the new properties of the transgenic plants, their supporters have valid commercial arguments to support their production, cultivation and trading.

4. Advantages of transgenic plants presented by their producers and supporters

The supporters of transgenic plants producing and trading say that these have a lot of advantages both for producers, for farmers, for industry, for consumers and for the environment and the human future.

a.   For the producers of the new varieties

A high efficiency in plants amelioration is obtained. The techniques of gene transfer are more precise because they allow the insulation and the propagation of the interest gene, while the classical hybridization techniques use the entire parental genomes and for this reason are needed back-crossings to emphasize the manifestation of a parental gene or to eliminate some secondary unwanted effects determined by the action of the gene in the genome. Furthermore, the number of the new characters susceptible to be conferred by gene transfer is much higher because the entire genetic information could be used despite its origin (viral, bacterial, vegetal or even human).

b.   For farmers

First of all, the process of pests destroying is simplified due to the elimination of herbicides in the pre-emergent period and in the vegetation period. For the GMO only one total herbicide is necessary.

On the other hand, the production output is increasing as well as the profits of the transgenic cultures, even the obtaining cost of the GMO is rather high.

c.   For industry

Due to the new properties of the transgenic plants, their processing could be also improved, as is the case of the modified starch, of low lignin content wood (in this case the paper manufacturing is less pollutant), of bio-plastics, of some human protein production (easier and in higher quantities, for therapeutic aim).

d.   For consumers

Nowadays the fruits and the vegetables with delayed maturation are easier stored, with minimum losses. The maturation moment can be controlled according to the demands of the market.

In the future it is considered that transgenic plants can determine an improved human health due to the higher content of vitamins, minerals, essential aminoacids, by using the vaccine plants, the rice enriched in pro-vitamin A, etc.

e.   For the environment and human future

First of all, transgenic plants imply lower pollution due to lower quantities of pesticides. Then, higher agricultural productions are obtained and people hope to eliminate the starving in the world (by extension of the areas cultivated with GMO resistant against salted soils, acid soils, lower temperatures, etc.)

5. Risks related to the cultivation of the transgenic plants

Despite of their advantages, it seems that the obtaining and mostly the cultivation of transgenic plants also imply some risks. The most important risks are the followings:

a.   Risks related to the nowadays techniques of vegetal gene transfer
* Secondary unwanted effects. The first obtained transgenic tomato with delayed maturation was floury, with metallic taste and difficult to be transported due to its very fragile skin. Due to these reasons, the American consumers rejected it.

- "weaknesses" in the transgene expression. For example, in USA, in 1996, the fields with transgenic cotton were destroyed in proportion of about 60% by some insects against which the plants were considered to be resistant. Similar, also in cotton, after the second treatment with herbicide it was noticed a deformation of the capsule. This could have happened because the producers have had no enough time to check the stability of the transgenic character on an enough great number of experimental fields.

b.   Ambiental risks
- Limitation of the risk evaluation by experimental cultures
- Risks related to the health, materialized by some allergies and resistance to some antibiotics.
- Risks related to the biodiversity of the ecosystems by
  - dissemination of the transgenic pollen to the similar spontaneous plants;
  - crossings between transgenic and conventional varieties of species;
  - apparition of some pest plants resistant against total herbicides
- Risks related to plants resistant to the insects attack:
  - apparition of some pests resistant against the insecticide-protein of the Bt maize;
  - intoxication of other insects by the transgenic plants
  - toxicity for the enemies of the pest insects.
- Risks regarding the circuit of the insecticide toxins in soils and in the trophic chains;
- Risks determined by the cultivation techniques of the plants resistant against herbicide, insects, viruses.
- Risk of destruction of the spontaneous flora and of the plants in the neighborhood of the cultivated fields, by the total herbicides.

6. Cultivation of the genetic modified plants worldwide

The first commercial transgenic plants cultures started in the middle of 1990 years, in USA, where they had the fastest evolution.

Immediately after followed Argentina and Canada, where the increasing of the surfaces cultivated with transgenic plants stabilized beginning with 1999. Both countries were followed by Brazil and China.

After only 7 years from their official start, the commercial cultures of transgenic plants reached a total surface of 60 million ha at which we could add the illegal GM soy cultures in Brazil of about 1 million ha.

During the fourteen years of commercialization 1996 to 2009, the global area of biotech crops increased almost 80-fold (78.8), from 1.7 million hectares in 1996 to 134 million hectares in 2009. This rate of adoption is the highest rate of crop technology adoption for any crop technology and reflects the continuing and growing acceptance of biotech crops by farmers in both large as well as small farms and resource-poor farmers in industrial and developing countries. In the same period, the number of countries growing biotech crops quadrupled , increasing from 6 in 1996 to 12 countries in 1999, 17 in 2004, 21 countries in 2005, and 25 in 2009 (Clive J., 2010).

Taking into account that approximately 21% of the 134 million hectares had two or three traits (planted primarily in the USA, but also increasingly in ten other countries, Argentina, Canada, the Philippines, South Africa, Australia, Mexico, Chile, Colombia, Honduras, and Costa Rica), the true global area of biotech crops in 2009 expressed as "trait hectares" was 180 million compared with 166 million "trait hectares" in 2008. Thus, the real growth rate measured in "trait hectares" between 2009 (180 million) and 2008 (166 million) was 8% or 14 million hectares compared with the apparent growth rate of 7% or 9 million hectares when measured conservatively in hectares between 2008 (125 million hectares) and 2009 (134 million hectares). (Clive J., 2010).

The list of the higher transgenic plants which are produced includes: (http://www.molecular-plant-biotechnology.info/transgenic-plants/list-of-higher-plants-where-transgenic-plants-have-been-produced.htm)

Herbaceous dicotyledons, such as: Nicotiana tabacum (tobacco), N. plumbaginifolia (wild tobacco), Petunia hybrida (petunia), Lycopersicon esculentum (tomato), Solanum tuberosum (potato), Solanum melongena (eggplant), Arabidopsis thaliana, Lactuca sativa (lettuce), Apium graveolens (celery), Helianthus annuus (sunflower), Linum usitatissimum (flax), Brassica napus (oilseed rape; canola), Brassica oleracea (cauliflower), Brassica oleracea var (cabbage), Brassica rapa (syn. B. campestris), Gossypium hirsutum (cotton), Beta vulgaris (sugarbeet), Glycine max (soybean), Pisum sativum (pea), Medicago sativa (alfalfa), M. varia, Lotus corniculatum (lotus), Vigna aconitifolia, Cucumis sativus (cucumber), Cucumis mew (muskmelon), Cichorium intybus (chicory), Daucus carota (carrot), Armoracia sp. (horse radish), Glycorrhiza glabra (licorice), Digitalis' purpurea (foxglove), Ipomoea batatas (sweet potato), Ipomoea purpurea (morning glory), Fragaria sp. (strawberry), Actinidia sp. (Kiwi), Carica papaya (papaya), Vitis vinifera (grape), Vaccinium macrocarpon (cranberry), Dianthus caryophyllus (carnation), Chrysallthemum sp. (chrysanthemum), Rosa sp. (rose);

Woody dicotyledons: Populus sp. (poplar), Malus sylvestris (apple), Pyrus communis (pear), Azadirachta indica (neem), Juglans regia (walnut);

Monocotyledons: Asparagus sp. (asparagus), Daclylis glomerata (orchard grass), Secale cereale (rye), Oryza sativa (rice), Triticum aestivum (wheat), Zea mays (corn), Avena sativa (oats), Festuca arundinacea (tall fescue);

Gymmosperms (a conifer) Picea glauca (white spruce).

Analysing the list it can be noticed that nowadays genetic engineering helped the producing of a wide scale of different types of plants but just a few of them are nowadays cultivated at large scale.

The worldwide market of transgenic plants consists almost exclusively, in four species: soy, cotton, maize and rapes. The other transgenic plants (potato, papaya, tobacco, pumpkin) are cultivated only on small surfaces, non-relevant for the total cultivated ones.

The list of the EU registered GM products which are used as food or food additives comprise the following plants:

(http://ec.europa.eu/food/dyna/gm_register/index_en.cfm)

-   Soybean (MON40-3-2), MON-Ø4Ø32-6, Monsanto, genetically modified soybean that contains: cp4 epsps gene inserted to confer tolerance to the herbicide glyphosate.
-   Soybean (A2704-12), ACS-GMØØ5-3, Bayer, genetically modified soybean that contains: pat gene inserted to confer tolerance to the glufosinate-ammonium herbicide.
-   Soybean (MON89788), MON-89788-1, Monsanto, genetically modified soybean that contains: cp4 epsps gene inserted to confer tolerance to the herbicide glyphosate.
-   Cotton (MON1445), MON-Ø1445-2, Monsanto, genetically modified cotton that contains: cp4 epsps gene inserted to confer tolerance to the herbicide glyphosate.
-   Cotton (MON15985) MON-15985-7, Monsanto, genetically modified cotton that contains: cry1Ac and cry2Ab2 genes inserted to confer insect-resistance highly selective in controlling Lepidopteran insects.
-   Cotton (MON15985 x MON1445), MON-15985-7 x MON-Ø1445-2, Monsanto, Genetically modified cotton that contains: cry1Ac and cry2Ab2 genes inserted to confer insect-resistance highly selective in controlling Lepidopteran insects and cp4 epsps gene inserted to confer tolerance to the herbicide glyphosate.
-   Cotton (MON531), MON-ØØ531-6, Monsanto, genetically modified cotton that contains: cry1A(c) gene inserted to confer insect-resistance.
-   Cotton (MON531 x MON1445), MON-ØØ531-6 x MON-Ø1445-2, Monsanto, Genetically modified cotton that contains: cry1A(c) gene inserted to confer insect-resistance and cp4 epsps gene inserted to confer tolerance to the herbicide glyphosate.
-   Cotton (LLCotton25), ACS-GHØØ1-3, Bayer, genetically modified cotton that contains: pat gene inserted to confer tolerance to the glufosinate-ammonium herbicide.
-   Cotton (GHB614), BCS-GHØØ2-5, Bayer, genetically modified cotton that expresses: 2mepsps gene inserted to confer tolerance to the glyphosate herbicides.
-   Maize (Bt11), SYN-BT Ø11-1, Syngenta, genetically modified maize that contains: the cry1A (b) gene inserted to confer insect-resistance and the pat gene inserted to confer tolerance to the herbicide glufosinate-ammonium.
-   Maize (DAS59122), DAS-59122-7, Pioneer and Dow AgroSciences, genetically modified maize that contains: the cry34Ab1 and cry35Ab1 genes inserted to confer protection against certain coleopteran pests such as corn rootworm larvae (Diabrotica spp.) and pat gene inserted to confer tolerance to the glufosinate-ammonium herbicide.
-   Maize (DAS1507), DAS-Ø15Ø7-1, Pioneer and Dow AgroSciences, genetically modified maize that contains: cry1F gene inserted to confer resistance to the European corn borer and certain other lepidopteran pests and pat gene inserted to confer tolerance to the herbicide glufosinate-ammonium.
-   Maize (DAS1507xNK603), DAS-Ø15Ø7-1xMON-ØØ6Ø3-6, Pioneer and Dow AgroSciences, genetically modified maize that contains: cry1F gene inserted to confer protection against certain lepidopteran pests such as the European corn borer (Ostrinia nubilalis) and species belonging to the genus Sesamia, pat gene inserted to confer tolerance to the glufosinate-ammonium herbicide and cp 4epsps gene inserted to confer tolerance to the glyphosate herbicide.
-   Maize (GA21), MON-ØØØ21-9, Syngenta, genetically modified maize that contains: mepsps gene inserted to confer tolerance to herbicide glyphosate.
-   Maize (MON810) , MON-ØØ81Ø-6, Monsanto, genetically modified maize that contains: cry1A (b) gene inserted to confer resistance to lepidopteran pests.
-   Maize (MON863), MON-ØØ863-5, Monsanto, genetically modified maize that contains: a trait gene cry3Bb1 inserted to confer insect- resistance and nptII gene inserted as a selection marker.

- Maize (NK603), MON-ØØ6Ø3-6, Monsanto, genetically modified maize that contains: cp4 epsps gene inserted to confer tolerance to the herbicide glyphosate.
- Maize (NK603 x MON810), MON-ØØ6Ø3-6 x MON-ØØ81Ø-6, Monsanto, genetically modified maize that contains: cp4 epsps gene inserted to confer tolerance to glyphosate herbicides and the cry1Ab gene inserted to confer protection against certain lepidopteran insect pests (Ostrinia nubilalis, Sesamia spp.).
- Maize (T25), ACS-ZMØØ3-2, Bayer, genetically modified maize that contains: pat gene inserted to confer tolerance to the herbicide glufosinate-ammonium.
- Maize (MON88017), MON-88Ø17-3, Monsanto, genetically modified maize that contains: modified cry3Bb1 gene inserted to confer protection to certain coleopteran pests and cp4 epsps gene inserted to confer tolerance to glyphosate herbicides.
- Maize (MON89034), MON-89Ø34-3, Monsanto, genetically modified maize that contains: cry1A.105 and cry2Ab2 genes inserted to confer protection to certain lepidopteran pests.
- Maize (59122xNK603), DAS-59122-7xMON-ØØ6Ø3-6, Pioneer, genetically modified maize that contains: cry34Ab1 and cry35Ab1 genes inserted to confer protection against certain coleopteran pests; pat genes inserted to confer tolerance to the glufosinate-ammonium herbicides and cp4 epsps genes inserted to confer tolerance to glyphosate herbicides.
- Maize (MIR604), SYN-IR6Ø4-5, Syngenta, genetically modified maize that contains: modified cry3A gene inserted to confer protection against certain coleopteran pests and pmi gene inserted as selection marker.
- Maize (MON863xMON810xNK603) , MON-ØØ863-5xMON-ØØ81Ø-6xMON-ØØ6Ø3-6, Monsanto, genetically modified maize that contains: cry3Bb1 gene inserted to confer protection against certain coleopteran pests; cry1Ab gene inserted to confer protection against certain lepidopteran insect pests; cp4 epsps gene inserted to confer tolerance to glyphosate herbicides and nptII gene inserted as a selection marker.
- Maize (MON863 x MON810) , MON-ØØ863-5 x MON-ØØ81Ø-6, Monsanto, genetically modified maize that contains: cry3Bb1 gene inserted to confer protection against certain coleopteran pests; cry1Ab gene inserted to confer protection against certain lepidopteran insect pests and nptII gene inserted as a selection marker.
- Maize (Bt11xGA21), SYN-BTØ11-1xMON-ØØØ21-9, Syngenta, genetically modified maize that expresses: the cry1Ab gene which confers protection against certain lepidopteran pests ; the pat gene which confers tolerance to the glufosinate-ammonium herbicides and the mepsps gene which confers tolerance to glyphosate herbicides.
- Maize (MON863 x NK603), MON-ØØ863-5 x MON-ØØ6Ø3-6, Monsanto, genetically modified maize that contains: cry3Bb1 gene inserted to confer protection against certain coleopteran pests; cp4 epsps gene inserted to confer tolerance to glyphosate herbicides and nptII gene inserted as a selection marker.
- Maize (MON88017xMON810), MON-88Ø17-3xMON-ØØ81Ø-6, Monsanto, genetically modified maize that expresses: the cry1Ab gene which confers protection against certain lepidopteran pests ; the cry3Bb1 gene which provides protection to certain coleopteran pests and the cp4 epsps gene which confers tolerance to glyphosate herbicides.
- Maize (MON89034 xNK603), MON-89Ø34-3x MON-ØØ6Ø3-6, , Monsanto, genetically modified maize that expresses: the cry1A.105 and cry2Ab2 genes which provide protection to certain lepidopteran pests and the cp4 epsps gene which confers tolerance to glyphosate herbicides.

- Maize (59122x1507xNK603), DAS-59122-7xDAS-Ø15Ø7xMON-ØØ6Ø3-6, Pioneer, genetically modified maize that expresses: the cry1F gene which confers protection against certain lepidopteran pests; the cry34Ab1 and cry35Ab1 genes which provide protection to certain coleopteran pests; the pat gene which confers tolerance to the glufosinate-ammonium herbicides and the cp4 epsps gene which confers tolerance to glyphosate herbicides.
- Maize (1507x59122), DAS-Ø15Ø7x DAS-59122-7, Pioneer, genetically modified maize that expresses: the cry1F gene which confers protection against certain lepidopteran pests; the cry34Ab1 and cry35Ab1 genes which provide protection to certain coleopteran pests and the pat gene which confers tolerance to the glufosinate-ammonium herbicides.
- Maize (MON89034 xMON88017), MON-89Ø34-3x MON-88Ø17-3, Monsanto, genetically modified maize that expresses: cry1A.105 and cry2Ab2 genes which provide protection to certain lepidopteran pests; cry3Bb1 gene which provides protection to certain coleopteran pests and cp4 epsps gene which confers tolerance to glyphosate herbicides.
- Oilseed rape (GT73), MON-ØØØ73-7, Monsanto, genetically modified oilseed rape that contains: cp4 epsps and goxv247 genes inserted to confer tolerance to the herbicide glyphosate.
- Oilseed rape (T45), ACS-BNØØ8-2, Bayer, genetically modified oilseed rape that contains: pat gene inserted to confer tolerance to the herbicide glufosinate-ammonium.
- Swede-rape (MS8, RF3, MS8xRF3), ACS-BNØØ5-8ACS-BNØØ3-6ACS-BNØØ5-8 x ACS-BN003-6, Bayer, genetically modifieds oilseed rape that contains:a bar (pat) gene inserted to confer tolerance to herbicides based on glufosinate ammonium; barnase gene inserted to leads to lack of viable pollen and male sterility and barstar gene inserted to leads to lack of viable pollen and male sterility.
- Starch potato (EH92-527-1), BPS-25271-9, BASF, genetically modified starch potato that contains: an inhibited gbss gene responsible for amylase biosynthesis. As a result, the starch product has little or no amylase and consists of amylopectin and nptII gene inserted as a selection marker.
- Sugar beet (H7-1), KM-ØØØ71-4, KWS SAAT and Monsanto, genetically modified sugar beet that expresses: a CP4 EPSPS protein confers tolerance to glyphosate containing herbicides.

Beside the GMO plants there also exist genetic modified microorganisms, such as: (http://ec.europa.eu/food/dyna/gm_register/index_en.cfm)

Bacterial biomass, (pCABL- Bacterial biomass), Ajinomoto Eurolysine SAS, Bacterial protein, by-product from the production by fermentation of L-Lysine HCl obtained from (Brevibacterium lactofermentum) the recovered killed microorganisms. The source is the Brevibacterium lactofermentum strain SO317/pCABL, used for feed produced from GMO bacteria: " bacterial biomass".

Yeast biomass, (pMT742 or pAK729-Yeast biomass), NOVO Nordisk A/S, NOVO Yeast Cream is a product produced from genetically modified yeast strains (Saccharomyces cerevisiae) cultivated on substrates of vegetable origin. The source is the Saccharomyces cerevisiae strain MT663/pMT742 or pAK729, used for feed materials produced from GMO yeast: "yeast biomass".

The global impact of the genetic modified plants presents the following aspects *(www.biotech-gmo.com)*: the additional brut margin realized by the farmers by cultivation of the genetic modified plants is of 22 billion USD, and the reduction of the pesticides quantities applied on the soil for the modified plants is about 172 million kg, resulting an impact coefficient on the environment of 14%.

Herbicide tolerance continues to be the most common transgenic trait. Herbicide tolerance is available for all of the major GM crops, including soybean, maize, rapeseed, and cotton. In 2005, the first herbicide tolerant sugar beets were approved in the US, Australia, Canada, and the Philippines. Herbicide tolerant rice and wheat already have been developed, but currently are not in use. In 2006, there was wide cultivation of herbicide tolerant alfalfa for the first time in the USA (80,000 hectares). In most of the cases the tolerance is to the following herbicides glyphosate *(Roundup)* or glufosinate-ammonium *(Liberty)*. Such crops make up 70 percent of the 102.0 million hectares of GM crops worldwide (2006).

(http://www.gmo-compass.org/eng/agri_biotechnology/gmo_planting/145.gmo_
cultivation_trait_statistics.html)

Global area of genetically engineered crops, 1996 to 2006: By trait (million hectares)					
Trait	HT	IR (Bt)	IR/HT	VR/Others	Total
1996	0.6	1.1	--	<0.1	1.7
1997	6.9	0.4	<0.1	<0.1	11.0
1998	19.8	7.7	0.3	<0.1	27.8
1999	28.1	8.9	2.9	<0.1	39.9
2000	32.7	8.3	3.2	<0.1	44.2
2001	40.6	7.8	4.2	<0.1	52.6
2002	44.2	10.1	4.4	<0.1	58.7
2003	49.7	12.2	5.8	<0.1	67.7
2004	58.6	15.6	6.8	<0.1	81.0
2005	63.7	16.2	10.	<0.1	90.0
2006	69.9	19.0	13.1	<0.1	102.0

Source: ISAAA, Clive James, 2006.

HT	Herbicide tolerance
IR	Insect resistance (mostly Bt)
VR	Resistance to virus diseases

Table 1. Global area of genetically engineered crops

Insect resistance is the second most common genetically modified trait. Herbicide tolerance and insect resistance (Bt) often are introduced simultaneously to a crop in one transformation event. This is called trait stacking. The third most commonly grown transgenic crop was stacked insect resistant/herbicide tolerant maize. Combined herbicide and insect resistance was the fastest growing GM trait from 2004 to 2005, grown on over 6.5 million hectares in the US and Canada and comprising seven percent of the global biotech area. The recent expansion of Bt crops is mainly due to the increasing Bt maize and Bt cotton production in China, India, and Australia (Table 1 and 2).

Trait-crop combinations, 1996 to 2005 (million hectares)						
Trait	IR maize	HT maize	IR/HT maize	IR/HT cotton	HT cotton	IR cotton
1996	0.3	0.0	--	0.0	< 0.1	0.8
1997	3.0	0.2	--	< 0.1	0.4	1.1
1998	7.0	2.0	--	--	--	1.0
1999	7.5	1.5	2.1	0.8	1.6	1.3
2000	6.8	2.1	1.4	1.7	2.1	1.5
2001	5.9	2.4	2.5	1.9	1.8	2.1
2002	7.7	2.5	2.2	2.2	2.2	2.4
2003	9.1	3.2	3.2	2.6	1.5	3.1
2004	11.2	4.3	3.8	3.0	1.5	4.5
2005	11.3	3.4	6.5	3.6	1.3	4.9
2006	--	--	--	--	--	3.8

Source: ISAAA, Clive James, 2006.

HT	Herbicide tolerance
IR	Insect resistance (mostly Bt)
VR	Resistance to virus diseases

Table 2. Trait –crop combinations in GMOs

7. Legislation regarding the transgenic plants

The purpose of a legislation system regarding the utilisation and cultivation of transgenic organisms obtained by modern biotechnology is the protection of the environment and the human health. The legislation has a preventive role and not a corrective one.

The legislation regulating the obtaining, testing, utilisation and commercialisation of the organisms obtained by modern biotechnology was elaborated in 1990 with three main objectives:

- to protect the human and animal health;
- to protect the environment;
- to assure the circulation of the GMO in EU;

The most important legislative act regarding GMO's in EU is the Directive 2001/18/EC and the Regulation (EC) No. 178/2002.

7.1 Directive 2001/18/EC

The directive presents the legislative framework regarding the deliberate release of the genetically modified organisms (GMOs) into the environment and the placing of GMOs on the market in accordance with the precautionary principle.

(http://europa.eu/legislation_summaries/agriculture/food/l28130_en.htm).

The main aim of this Directive is to make the procedure for release and placing on the market of genetically modified organisms (GMOs) and to introduce compulsory monitoring after GMOs have been placed on the market.

It also provides a common methodology to assess case-by-case the risks for the environment associated with the release of GMOs (the principles applying to environmental risk assessment are set out in Annex II to the Directive).

Public consultation and GMO labelling are made compulsory under the new Directive. Rules on the operation of these registers are laid down in Decision 2004/204/EC

The Directive invited the Commission to present a proposal for implementing the Cartagena Protocol on biosafety, which led to the adoption of Regulation (EC) No 1946/2003 on transboundary movements of genetically modified organisms.

The Directive 2001/18/EC had entry into force in 17.4.2001, the deadline for transposition in the member states of the EU was 17.10.2002 and the Official Journal in which it was published was OJ L 106 of 17.4.2001. The amending acts of Directive 2001/18/EC are represented by two regulations, Regulation (EC) No 1829/2003 and Regulation (EC) No 1830/2003, both entering into force starting with 07.11.2003, and being published into OJ L 268 of 18.10.2003.

(http://europa.eu/legislation_summaries/agriculture/food/l28130_en.html).

7.2 Regulation (EC) No. 178/2002

Regulation (EC) No. 178/2002 of the European Parliament and of the Council of 28 January 2002 laying down the general principles and requirements of food law, establishing the European Food Safety Authority (EFSA) and laying down procedures in matters of food safety.

This Regulation provides a framework for food and feed Law within the EC. It applies to all stages of production, processing and distribution of food and feed, but does not apply to primary production for private domestic use or to the domestic preparation, handling or storage of food for private domestic consumption.

(http://www.food.gov.uk/scotland/regsscotland/regulations/scotlandfoodlawguide/sflg 200501/).

The legislation regarding the authorization for introduction of the GMO's in the environment for the experimental purpose is presented in the Directive 2001/18/EC (part B).

Important information, regulated in the part C, are about the general aspects of GMO's labelling, free circulation and information to the public.

(http://www.biosafety.be/GB/Dir.Eur.GB/Del.Rel./2001_18/2001_18_TC.html)

7.3 Other legislation regarding GMO's

Other legal instruments regarding GMO's are the followings:

- Directive 90/220/EEC on the deliberate release into the environment of genetically modified organisms entry into force on the 17-th of October 2002;
- Directive 98/81/EC amending Directive 90/219/EEC regarding the use of GMO's (http://europa.eu.int/eur-lex/pri/en/oj/dat/1998/1_330/1_33019981205en00130031.pdf)
- Decision 2002/623/EC establishing guidance notes supplementing Annex II to Directive 2001/18/EC. (http://europa.eu.int/eur-lex/pri/en/oj/dat/2002/1_200/1_20020020730en00220033.pdf)
- Decision 2002/813/EC115 establishes the format to be used by competent authorities when they provide summaries of notifications to the Commission under Article 11 of Directive 2001/18/EC. (http://europa.eu.int/eur-lex/pri/en/oj/dat/2002/1_280/1_28020021018en00620083.pdf)
- Decision 2003/701/EC116 establishes the format to be used by notifiers in the reporting of the results of the deliberate release to the competent authorities, as required by Article 10 of Directive 2001/18/EC (http://europa.eu.int/eur-lex/pri/en/oj/dat/2003/1_254/1_25420031008en00210028.pdf)
- Decision 2002/812/EC establishing pursuant to Directive 2001/18/EC the summary information format relating to the placing on the market of genetically modified organisms as or in products. (http://europa.eu.int/eur-lex/pri/en/oj/dat/2002/1_280/1_28020021018en00370061.pdf)
- Decision 2004/204/EC laying down detailed arrangements for the operation of the registers for recording information on genetic modifications in GMO, provided in Directive 2001/18/EC. (http://europa.eu.int/eur-lex/pri/en/oj/dat/2002/1_280/1_28020021018en00270036.pdf)
- Decision 2002/811/EC establishing guidance notes supplementing Annex II to Directive 2001/18/EC. (http://europa.eu.int/eur-lex/pri/en/oj/dat/2002/1_200/1_20020020730en00220033.pdf)
- Regulation (EC) No. 178/2002 regarding the general principles and requirements of food law. (http://europa.eu.int/eur-lex/pri/en/oj/dat/2002/1_031/1_03120020201en00010024.pdf)
- Regulation (EC) No. 1829/2003 on genetically modified food and feed; (http://europa.eu.int/eur-lex/pri/en/oj/dat/2003/1_268/1_26820031018en00010023.pdf)
- Regulation (EC) No. 1830/2003 concerning the traceability and labelling of genetically modified organisms and traceability of food and feed products produced from genetically modified organisms. (http://europa.eu.int/eur-lex/pri/en/oj/dat/2003/1_268/1_26820031018en00240028.pdf)

- Regulation (EC) No. 1946/2003 on transboundary movements of genetically modified organisms.
(http://europa.eu.int/eur-lex/pri/en/oj/dat/2003/1_287/1_28720031105en00010010.pdf)

8. Genetical modified flowers on the market

The research regarding the color of the flower is an ongoing process. It was noticed that in some plants the blue pigment is lacking, so the blue colored flowers cannot be obtained by classical cross-breed techniques. Genetic engineering has allowed scientists to produce carnation blue flowers even that the blue pigment is missing.

Moondust' carnation, first grown commercially in 1997, is a mini-carnation with purple-mauve flowers that gets its blue color from petunia genes grafted into the DNA of the carnation. Twelve scientists at an Australian company called Florigene labored for a decade to isolate the gene responsible for blue color in petunia and then transfer it into the carnation.

To date, they have released five carnations with the "Moon" prefix, all with varying shades of mauve (Moonvista), blue (Moonshade), violet (Moonlite) or purple (Monaqua) (Fig.1).

Roses, carnations, lilies and orchids all lack a class of blue pigments called delphinidins, named after the violet-blue we see in delphinium. The gene for delphinidin production is what the Florigene scientists removed from petunia and transferred to the carnation. (http://www.arhomeandgarden.org/plantoftheweek/articles/blue_carnation.htm)

| Moonvista | Moonshade | Moonlite | Moonaqua |

Fig. 1. Florigene blue carnations

The blue carnations are cultivated in Ecuador, Columbia and Australia and commercialised in USA, Canada, Japan, EU and Australia.

Blue roses were always very desirable flowers because they symbol the mystery, the untouchable, being impossible to be produced free in nature. They were avaible on many markets, from many years, but they were not real blue roses, because they are traditionally created by dyeing white roses with blue dye.

It was in 2004 when the first blue rose was obtained by genetic engineering (Fig. 2). It was the result of the researches made by Japanese (Suntory Limited Research Centre) and Australian (Florigene) researchers and the first blue roses were officially presented to the specialists and to the market at the World Rose Convention in Osaka 2006.

http://www.flowermeaning.info/Blue.php

Fig. 2. Blue rose presented at World Rose Convention in Osaka, 2006

The above mentioned researchers observed that the flower color is mainly determined by anthocyanins. *Rosa hybrida* lacks violet to blue flower varieties due to the absence of delphinidin-based anthocyanins, usually the major constituents of violet and blue flowers, because roses do not possess flavonoid 3',5'-hydroxylase (F3'5'H), a key enzyme for delphinidin biosynthesis. Other factors such as the presence of co-pigments and the vacuolar pH also affect flower color.

It was analyzed the flavonoid composition of hundreds of rose cultivars and measured the pH of their petal juice in order to select hosts of genetic transformation that would be suitable for the exclusive accumulation of delphinidin and the resulting color change toward blue. Expression of the viola *F3'5'H* gene in some of the selected cultivars resulted in the accumulation of a high percentage of delphinidin (up to 95%) and a novel bluish flower color.

For more exclusive and dominant accumulation of delphinidin irrespective of the hosts, the researchers down-regulated the endogenous dihydroflavonol 4-reductase (*DFR*) gene and over-expressed the *Iris×hollandica DFR* gene in addition to the viola *F3'5'H* gene in a rose cultivar. The resultant roses exclusively accumulated delphinidin in the petals, and the flowers had blue hues not achieved by hybridization breeding. Moreover, the ability for exclusive accumulation of delphinidin was inherited by the next generations. (Yukihisa Katsumoto, 2007).

9. Conclusions

1. The scientific progress could be not stopped by anyone and by nothing. It is very important to use the scientific progress for the whole humanity and to be in

concordance with the actual and the future generations' interest, without supporting the interests of only certain groups. This was the starting point of sustainability strategy which, in its essence, stipulate that the development is made for people and is realized by people.

2. The ecologic, geographic and antropic assemblies realize the landscape, with very important functions of general interest at the cultural, ecological, social level, this landscape being an important resource of the human economic activities. In this context, it should be mentioned that we don't know yet all the potential risks that the GMO could have, by long term accumulation, upon the environment.

3. The achievements of the genetic engineering have nowadays considerable benefits, but now we don't know the price we, or the future generations, will have to pay in the future for this benefits. The long term risks of the GMO are not entirely known today.

10. References

Apostu, S. (2006). Organisme modificate genetic folosite în alimentaţia şi sănătatea omului şi animalelor, Ed. Risoprint, Cluj-Napoca, ISBN 973-606-504-084-7.

Badea E. & Otiman IP (2006). Plante modificate genetic în cultură, Ed. Mirton, Timisoara, ISBN 973-8266-36-X.

Brookes G & P.Barfoot (2008). GM crops: global socio-economic and environmental impact 1009-2006, http://www.pgeconomics.co.uk/pdf.

Cornea C.P.; I. Vătafu & A. Barbu (1998). Elemente de inginerie genetică, Ed. All Educaţional, Bucureşti,), ISBN 973-647-282-5.

Cristea S. & Simone Denaeyer (2004). De la biodiversitate la OGM-uri? , colecţia Universitas, seria Biologie, Ed. Eikon, ISBN 973-656-036-8.

Desmond N. (2002). Genetic Engineering, Cambridge University Press.

James C. (2005). Global Status of Commercialized Biotech/GM Crops: 2009. ISAAA Brief No.39.

James C. (2008). Global Status of Commercialized Biotech/GM Crops: 2009. ISAAA Brief No.34.

James C. (2009). Global Status of Commercialized Biotech/GM Crops: 2009. ISAAA Brief No. 41. ISAAA: Ithaca, NY.

Johnson T. (1999). Gene-Changed Foods Hit Local Market, Bussiness Week, Vol. 3. No. 35.

Malschi Dana (2007). Puncte de vedere asupra prezenţei organismelor genetic modificate (OGM) în agricultură, Revista Bioterra, ian-feb.-martie, 2007, p. 11-15.

Vlaic A. (1997), Inginerie genetică, Ed. Promedia Plus, Cluj-Napoca, 1997, ISBN 973-9204-71-6.

Yukihisa Katsumoto and col, 2007, Engineering of the Rose Flavonoid Biosynthetic Pathway Successfully Generated Blue-Hued Flowers Accumulating Delphinidin, Plant and Cell Physiology, Vol.48, Issue 11, pp. 1589-1600.

Consulted WEB sites:

www.biotech-gmo.com

http://www.aabiotech.org/resources

http://users.rcn.com/jkimball.ma.ultranet/BiologyPages/T/TransgenicPlants.html

http://www.gmo-compass.org/eng/agri_biotechnology/gmo_planting/
    145.gmo_cultivation_trait_statistics.html

http://www.isaaa.org/resources/publications/briefs/41/download/isaaa-brief-41-
    2009.pdf
http://www.biosafety.be/GB/Dir.Eur.GB/Del.Rel./2001_18/2001_18_TC.html
http://www.flowermeaning.info/Blue.php
http://europa.eu/legislation_summaries/agriculture/food/l28130_en.html
http://www.food.gov.uk/scotland/regsscotland/regulations/scotlandfoodlawguide/sflg2
    00501/
http://www.arhomeandgarden.org/plantoftheweek/articles/blue_carnation.htm
Convention on Biological Diversity, 2005, http://bch.biodiv.org/
http://europa.eu.int/eur-lex/pri/en/oj/dat/1998/l_330/l_33019981205en00130031.pdf
http://europa.eu.int/eur-lex/pri/en/oj/dat/2002/l_200/l_20020020730en00220033.pdf
http://europa.eu.int/eur-lex/pri/en/oj/dat/2002/l_280/l_28020021018en00620083.pdf
http://europa.eu.int/eur-lex/pri/en/oj/dat/2003/l_254/l_25420031008en00210028.pdf
http://europa.eu.int/eur-lex/pri/en/oj/dat/2002/l_280/l_28020021018en00370061.pdf
http://europa.eu.int/eur-lex/pri/en/oj/dat/2002/l_280/l_28020021018en00270036.pdf
http://europa.eu.int/eur-lex/pri/en/oj/dat/2002/l_200/l_20020020730en00220033 .pdf
http://europa.eu.int/eur-lex/pri/en/oj/dat/2002/l_031/l_03120020201en00010024.pdf
http://europa.eu.int/eur-lex/pri/en/oj/dat/2003/l_268/l_26820031018en00010023.pdf
http://europa.eu.int/eur-lex/pri/en/oj/dat/2003/l_268/l_26820031018en00240028.pdf)
http://europa.eu.int/eur-lex/pri/en/oj/dat/2003/l_287/l_28720031105en00010010.pdf)

10

Elimination of Transgenic Sequences in Plants by *Cre* Gene Expression

Lilya Kopertekh and Joachim Schiemann
*Julius Kuehn Institute, Federal Research Centre
for Cultivated Plants (JKI),
Institute for Biosafety of Genetically
Modified Plants, Quedlinburg,
Germany*

1. Introduction

The ability to insert foreign DNA into plant cells opened plenty opportunities for the development of new cell lines and improved varieties for agronomic and industrial purposes. Despite the great advances reached there are still some limitations in plant biotechnology based on genetic transformation. In most cases precise engineering of target genomic loci is difficult. Random DNA integration and multi-copy transgene insertions might result in unpredictable expression or gene silencing. Furthermore, commercial application of plant biotechnology products rises numerous regulatory and biosafety concerns about possible spread of the transgenes into the environment or the presence of selectable marker genes. One of the molecular tools that can help to overcome these limitations is site-specific recombination. Several site-specific recombination systems have been shown to be functional in plant cells: the Cre-*lox* system from bactreiophage P1 (Dale and Ow, 1990; Odell et al., 1990, Bayley et al., 1992), the FLP-*FRT* system from *Saccharomyces cerevisiae* (Lyznik et al., 1993; Lloyd and Davis, 1994; Kilby et al., 1995), the R-*RS* system from *Zygosaccharomyces rouxii* (Onouchi et al., 1991), the Gin-*gix* system from bacteriophage Mu (Maeser and Kahmann, 1991), the CinH-RS2 system from *Acetinetobacter* (Moon et al., 2011), the ParA system from a plasmid operon parCBA (Thomson et al., 2009) and the *Streptomyces* phage phiC31 system (Kittiwongwattana et al., 2007, Rubtsova et al., 2008). Currently, Cre-*lox* has become the most commonly employed site-specific recombination system. Although both types of recombination catalyzed by the Cre protein, site-specific integration and excision, found practical application (Ow, 2002; Gilbertson, 2003; Lyznik et al., 2003; Gidoni et al., 2008; Wang et al., 2011), the removal of *lox*-flanked sequences is the most widely used applications of Cre recombinase. The following technologies are based on excisional recombination: (i) regulation of gene expression, (ii) resolution of complex insertion sites to single copy structures, (iii) biological confinement, and (iv) elimination of selectable marker genes. Here we review the progress in the employment of Cre-mediated site-specific excisional recombination for applied plant biology and discuss in detail the advantages, limitations and potential improvements of technologies utilizing the Cre-*lox* system.

2. The Cre-*lox* site-specific recombination system: Structure, biological functions, mode of action

The Cre-*lox* site-specific recombination system from bacteriophage P1 belongs to the tyrosine integrase family whose members use a conserved tyrosine residue as catalytic nucleophile (Grindley et al., 2006). It performs at least two functions in the P1 life cycle: (i) it promotes the circularization of bacteriophage DNA after infection of bacteria (Segev and Cohen, 1981; Hochman et al., 1983), and (ii) it maintains the phage genome as unit-copy plasmid by resolving dimeric plasmids during bacterial division (Austin et al., 1981).

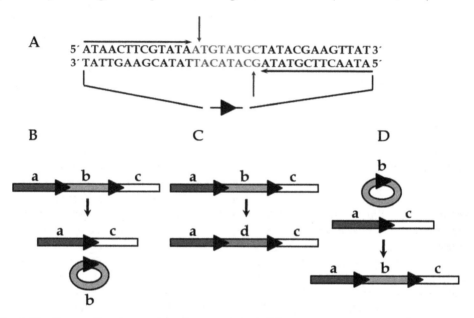

Fig. 1. The Cre-mediated recombination reaction. A: Schematic representation of the *lox* recombination site. The 13 bp inverted repeats are marked by large horizontal arrows. The points of the spacer region at which Cre cleaves the *lox* sites are denoted by small vertical arrows. Cre recombinase mediates inter- and intramolecular recombination leading to deletion (B), inversion (C) or integration (D) events.

The Cre-*lox* system consists of two short DNA recognition sequences known as *lox* (*locus of crossing-over*) and the recombinase protein Cre. Structural studies have revealed that a functional *lox* site is composed of two 13 bp inverted repeats flanking an 8 bp spacer region (Hoess et al., 1982; Hoess and Abremski, 1984) (Figure 1A). The inverted repeats and adjacing 4 bp of the spacer region compose a Cre binding domain. The asymmetry of the 8 bp spacer sequence determines the outcome of the recombination. The second component of the system, the 38 kDa Cre protein, includes two domains: a NH_2-terminal domain and a larger COOH-terminal domain, which contains the active site of the enzyme and major determinants for DNA binding specificity. The Cre recombinase does not require additional proteins or cofactors and performs enzymatic activity under a wide variety of cellular and non-cellular conditions. Crystallographic analysis of Cre-DNA complexes (Guo et al., 1997;

Guo et al., 1999) has revealed the recombination mechanisms. The process of site-specific recombination involves the formation and resolution of a Holliday junction intermediate, during which the DNA is transiently attached to the enzyme through a phosphotyrosine linkage. The reaction can result in integration, inversion or excision, depending on the position and orientation of the recombination sites. Recombination between two *lox* sites in direct orientation on the same DNA molecule results in excision of the *lox*-flanked DNA fragment (Figure 1B). In contrast, recombination between two *lox* sites in inverted repeat leads to inversion of the intervening DNA fragment (Figure 1C). Integration results from recombination between two *lox* sites situated on different DNA molecules (Figure 1D). The recombination reaction is reversible. Since intramolecular excision is kinetically favoured over bi-molecular integration, the excision reaction is essentially irreversible. In contrast, the insertion products are unstable in the presence of Cre recombinase.

3. Cre expression strategies: Efficiency and limitations

According to the presence of the *cre* sequence in the plant genome and the duration of *cre* expression, approaches to combine the *lox* recognition sequences and Cre protein can be grouped into three categories: (i) constitutive, (ii) transient and (iii) temporal expression. In the first group, the recombinase gene is stably integrated into the plant genome and expressed during the whole plant life. There are at least two main possibilities to integrate the *cre* gene into *lox*-containing plants: cross pollination and retransformation. To follow the crossing strategy, *cre* and *lox*-transgenic lines are developed and subsequently crossed (Bayley et al., 1992; Russell et al., 1992; Hoa et al., 2002). Applying the retransformation strategy, the *cre* gene is transformed into *lox*-lines (Odell et al., 1990; Dale and Ow, 1991; Zhang et al., 2003). Constitutive expression provides high recombination efficiencies in both model (Dale and Ow, 1991; Russell et al., 1992) and commercial crops (Hoa et al., 2002; Zhang et al., 2003). However, prolonged *cre* expression has some limitations. It is not optimal for plant species that are propagated by vegetative cuttings, since the crossing/segregation step for the *cre* gene can be problematic. Furthermore, additional time is required to perform a second round of transformation or cross pollination. A further strong argument against constitutive *cre* expression is the possible occurrence of genetic and phenotypic changes caused by the Cre recombinase, which were observed in plastid and nuclear genomes, respectively (Hajdukiewicz et al., 2001; Coppoolse et al., 2003).

Transient expression offers the possibility to reduce/avoid undesired side-effects caused by long-term persistence of the Cre protein. The following approaches have been described in the literature: application of the purified Cre protein and virus- or *Agrobacterium tumefaciens*-mediated *cre* expression.

Addition of Cre protein to induce site-specific recombination was initially demonstrated for animal cells (Baubonis and Sauer, 1993) and extended by Cao and co-workers (2006) to excise *lox*-flanked DNA fragments in plant culture. In theory, direct introduction of the recombinase protein into plant cells could be an elegant solution. In fact, the broad application of this method to commercial crops is highly problematic. Additional time and costs have to be invested to purify an enzymatically active Cre protein and to obtain optimal conditions for cell culture treatment. Reliable regeneration protocols from protoplasts are not available for several crops. Moreover, this regeneration step can introduce additional somaclonal variation.

The Cre function can be provided transiently by *Agrobacterium*-based vectors using T-DNA-independent and T-DNA-dependent expression. T-DNA-independent *Agrobacterium*-mediated *cre* expression is based on fusion of the Cre protein to the NH_2-terminus of VirE2 and VirF proteins. *Agrobacterium* is able to transfer these fusions into *Arabidopsis* cells resulting in excision events, although detectable efficiency of the process was low (Vergunst et al., 2000). Therefore, this system might be used only for applications where rare recombination rates are essential. T-DNA dependent expression relies on the fact that non-integrated copies of T-DNA may persist in the nucleus for a period of time providing transient expression of genes from T-DNA. The Cre recombinase gene cloned between left and right T-DNA borders can be delivered into plant cells by the agro-inoculation technique and recombine *lox* sites in both nuclear (Gleave et al., 1999; Kopertekh and Schiemann, 2005) and plastid (Lutz et al., 2006) genomes as shown in tobacco. The principle of transient recombinase expression via *A. tumefaciens*-based vectors was proved only in model plant species yet.

Another possibility to deliver Cre protein without *cre* gene insertion into the plant genome is provided by the application of RNA viruses. Two Cre-virus vectors, PVX-Cre (Kopertekh et al., 2004a, 2004b) and TMV-Cre (Jia et al., 2006), have been shown to be functional in *lox*-target *N. benthamiana* and *N. tabacum* plants. In both vectors the *cre* gene was integrated between movement and coat protein genes. Recently, the application of PVX-Cre for marker gene elimination in potato has been demonstrated (Kopertekh et al., 2011). In comparison to the *A. tumefaciens* transient expression system, virus vectors were more efficient in generating recombination events. In general, *Agrobacterium*- and virus-based *cre* expression is mostly suitable for vegetatively propagated species. However, the necessity to develop efficient agroinfiltration methods or infectious Cre-virus vectors, as well as regeneration protocols for plant explants might hamper a broad application of these approaches.

To follow the temporal expression approach, a stably integrated *cre* gene is placed under the control of inducible or tissue specific promoters. To date, a regulated *cre* expression is usually combined with the autoexcision strategy. Self-excision plant transformation vectors contain two recognition sites and the *cre* gene on the same T-DNA molecule. Conditional expression of the *cre* gene results in simultaneous removal of all sequences situated between the *lox* sites. This autoexcision strategy provides several potential advantages. First, all components of the Cre-*lox* system can be incorporated into the plant genome in one transformation step. Second, this strategy could be employed for both generatively and vegetatively propagated species. Several inducible systems responsive to external stimuli have been reported for plants, e.g. heat-shock and β-estradiol regulated. The heat-shock regulated system seems to be the simplest and most familiar for use. Its function has been demonstrated as functional in *Arabidopsis* (Hoff et al., 2001), tobacco (Wang et al., 2005), potato (Cuellar et al., 2006), maize (Zhang et al., 2003), rice (Khattri et al., 2011) and aspen (Fladung and Becker, 2010). In the chemically regulated self-excision system developed by Zuo and associates (2001), the *cre* gene was combined with the XVE system which is induced by β-estradiol. The system was successfully applied to *Arabidopsis* (Zuo et al., 2001), rice (Sreekala et al., 2005) and tomato (Zhang et al., 2006; Zhang et al., 2009). Despite the great advantage of the temporally controlled recombinase expression, heat-shock and chemically regulated promoters require an external signal to be activated and the recombination frequencies are greatly dependent on the penetration of the signal into plant cells, respectively.

A promising alternative to the *cre* regulation described above is the use of developmentally inducible promoters. During the last few years a number of promoters active in different stages of plant development, namely in germline (Verweire et al., 2007; Van Ex et al., 2009), embryo (Li et al., 2007), microspore (Mlynarova et al., 2006; Luo et al., 2007), floral (Bai et al., 2008) and seed (Odell et al., 1994; Moravčíková, et al., 2008; Kopertekh et al., 2010) tissues have been tested to control *cre* expression. High efficiency of such promoters in *Arabidopsis* (Verweire et al., 2007), tobacco (Mlynarova et al., 2006), rice (Bai et al., 2008), soybean (Lie et al., 2007) and oilseed rape (Kopertekh et al., 2009) makes this approach universal for model and agronomically important species. In addition, the employment of germline-specific promoters allows a more efficient transmission of the recombined status to the progeny. The essential feature of conditional Cre systems is a careful regulation with respect to time and tissue. Background Cre activation was observed for heat-shock inducible (Hoff et al., 2001; Wang et al., 2005) and some seed-specific promoters (Odell et al., 1994; Moravčíková, et al., 2008), resulting in reduced efficiency of the systems.

In summary, methodological progress in *cre* gene expression strategies allows to modulate the recombinase activity in a temporal manner. The choice between the Cre expression systems depends mainly on the goals of the experiment, involved plant species, and finally available expertise.

4. Application of Cre-mediated excision in plant biotechnology

The removal of *lox*-flanked DNA fragments by Cre recombinase is broadly used in plant applied research. The applications described in the literature can be grouped into four categories: (i) regulation of gene activity, (ii) simplification of complex transgene structures, (iii) complete excision of a transgene to prevent gene flow, and (iv) marker gene removal.

4.1 Gene regulation

Cre-mediated site-specific recombination offers an effective way to turn on or off gene expression in transgenic plants by removing DNA fragments located between directly repeated recombination sites. What are the potential uses of this technology?

One example is the use of plants as bioreactors to produce recombinant proteins that are toxic to plant cells. Tremblay et al. (2007) designed transgenic *Arabidopsis* plants harbouring a *Turnip Mosaic Virus* (TuMV) amplicon in which a *lox*-flanked translational terminator integrated between the P1 and HCPro coding sequences prevented virus replication. After delivery of Cre recombinase by agroinfiltration, a PVX-Cre vector or a transgenic chemically inducible system, the intervening DNA fragment was eliminated resulting in virus accumulation.

The same strategy was used for conditional recombinase-mediated gene expression in plant cell culture (Joubes et al., 2004). In a plant transformation vector, excision of the *gfp* coding sequence by heat-shock and a dexamethasone inducible Cre recombinase lead to expression of the gene of interest. The system was tested in *N. tabacum* bright yellow-2 (B-2) cells and its efficiency was demonstrated for the *gus* reporter gene and a potent inhibitor of the cell cycle mutant allele of the A-type cyclin-dependent kinase (CDKA).

Another example of recombinase-mediated gene regulation is the restoration of pollen fertility. Transgenic tobacco plants containing a *lox*-flanked stilbene synthase (*sts*) gene under control of a tapetum-specific promoter displayed the male-sterile phenotype (Bayer and Hess, 2005). Pollen fertility was restored after crossing with *cre*-expressing tobacco lines. This method may provide a valuable strategy for the production of hybrid plants.

In contrast to animal systems the few reports describing Cre recombinase-mediated gene regulation in plant systems only demonstrate a proof of principle without practical application yet.

4.2 Generation of single copy transformants by Cre-*lox* recombination

During plant genetic transformation multiple T-DNA copies are often integrated at a single locus. Complex integration sites are commonly associated with intrachromosomal recombination (Srivastava et al., 1996) and transgene silencing (Wang and Waterhouse, 2000; De Buck et al., 2001). Moreover, a single integration pattern may simplify the functional and structural characterization of a transgene. Therefore, single copy transgenic plants are more desirable for commercial practice. Several approaches such as conventional screening amongst a large pool of transformants (De Buck and Depicker, 2004), agrolistics (Hansen and Chilton, 1996), niacinamide application (De Block et al., 1997) or use of Cre-mediated site-specific recombination (Srivastava et al., 1999) have been developed to select/generate single copy lines. The Cre-*lox*-based strategy is based on a transgene flanked by *lox* sites in opposite orientation. In case of tandem insertion of T-DNAs at a single locus, the Cre recombinase resolves multiple units to a single-copy insert.

The proof of concept and successful application of the Cre-*lox*-based strategy was reported for the first time by Srivastava et al. (1999). Four transgenic wheat *lox*-target lines, containing a DNA fragment flanked by recombination sites in inverted repeats, were generated by particle bombardment. The Cre recombinase was provided by crossing with *cre*-expressing plants. Cre-mediated resolution of the complex T-DNA structure was observed in T_2 progeny plants for all four lines investigated. However the authors reported (i) incomplete resolution of complex loci in 20-40% of the T_2 progenies from three lines and (ii) persistence of excised DNA fragment extrachromosomally in one plant.

The strategy described above was modified to generate single-copy maize plants more efficiently. In comparison to the original method, the *cre*-expressing construct was introduced into *lox*-transgenic maize cells transiently by particle bombardment (Srivastava and Ow, 2001). This modification was highly efficient: 85% of regenerated plants contained 1 to 2 copies of the introduced DNA, with 38% harbouring a single copy. In 23% of single copy lines recombination was performed by transient *cre* expression: they harboured only the *lox*-target construct.

The Cre-mediated resolution approach was also functional in *Arabidopsis*. In the *lox*-transformation vector two recombination sites in inverted repeat were cloned inside the T-DNA immediately adjacent to the left and right T-DNA border ends (De Buck et al., 2007). Seven transgenic lines with a complex integration locus were crossed with *cre*-transgenic plants. The progeny of two hybrids demonstrated a single-copy T-DNA status without integration of the released DNA fragment in the plant genome. In some transformants, the Cre-mediated resolution of complex loci increased the transgene expression at least tenfold.

Based on these results an alternative transformation system to generate single copy transformants has been developed and proved in *Arabidopsis* (De Pape et al., 2009). To omit the crossing step between *cre*- and *lox*-plants, a *lox*-target construct was transferred by floral deep transformation into *cre* expressing plants. 55% of primary transformants contained a single copy of the introduced T-DNA. However 73% showed inversion of the DNA fragment between the *lox* sites which can result in variable transgene expression. Further improvement was achieved by introducing only one *lox* site in the transformation vector: 70% of primary transformants harboured a single-copy of T–DNA without inversion.

In summary, the recombinase-based resolution strategy can efficiently resolve complex integration patterns in important agricultural crops, particularly wheat and maize, as well as in the model plant *A. thaliana*. However, the following potential limitations have to be envisaged for this strategy. First, this approach may not be suitable for multiple locus integration events since Cre-mediated resolution can cause chromosomal deletions. Second, incomplete resolution of the complex locus is possible. Finally, released DNA fragment may be present in the plant genome.

4.3 Transgene confinement

One concern related to genetically modified plants is the potential effects resulting from transgene transfer into the environment. To address this issue several biological confinement strategies have been proposed. Current technologies, namely male sterility, chloroplast transformation, cleistogamy and transgene removal from pollen or seeds, offer new possibilities for biological confinement (Daniell, 2002; Keenan and Stemmer, 2002; Moon et al., 2009). In this chapter we will mainly describe biological confinement strategies based on the Cre-*lox* recombination system. Here, all functional transgenes are flanked by two recognition sites in direct orientation. Upon expression of the *cre* gene driven by tightly regulated chemically induced or tissue specific promoters, the transgene sequences are removed leaving only a short recognition sequence in the genome. Since gene flow occurs most frequently via seed or pollen dispersal, transgene removal from seed or pollen by developmentally regulated *cre* recombinase could minimize transgene transfer.

The seed-sterile technology is based on two expression units: *cre*-expression unit and cytotoxic ribosome-inhibitor (RIP) gene expression unit (Daniell, 2002). The *cre* gene is linked with a repressor-operator (Tet) system which allows *cre* expression in the presence of tetracycline. In the second expression unit, a seed-specific late embryonic abundance (LEA) promoter and a *RIP* gene are separated by a *lox*-flanked "spacer sequence". Tetracycline induced *cre* expression results in the removal of the "spacer sequence" and the fusion of LEA promoter and *RIP* gene. The RIP protein destructs the seed tissue resulting in production of non-viable seeds. The following potential problems are linked with this strategy: (i) all three components of the system (Cre, RIP and Tet) should be present together in one plant, (ii) the repressor-operator (Tet) system should display high efficiency in crop plants and the chemical inducer should penetrate the plant tissue uniformly, (iii) the seed-specific LEA promoter can be subjected to silencing causing undesired transgene dispersal.

In the second advanced strategy developed by Mlynarova et al. (2006), a *lox*-embedded cassette includes (i) marker gene, (ii) gene of interest and (iii) *cre* gene driven by the NTM 19 microspore-specific promoter. This design allows autoexcision of all transgenes during microsporogenesis without application of an additional induction factor. It was highly efficient in tobacco plants: only two out of 16800 seeds (0.024%) contained non-excised transgene sequences. Additionally, the authors did not observe premature activation or absence of activation for the tissue-specific Cre-system under laboratory stress conditions.

The efficiency and reliability of recombinase-mediated confinement methods was further improved by the application of pollen- and seed-specific promoters and hybrid *lox-FRT* recombination sites (Luo et al., 2007). The *lox-FRT* fusion sequences dramatically enhanced the excision frequency: analysis of 25000 progeny seedlings for several transgenic tobacco lines revealed that transgenes in pollen or seeds were excised with 100% efficiency. Despite simplicity and high efficiency of the developmentally regulated Cre-system to prevent gene flow, the need to maintain the hemizygous status may be a great disadvantage for transgenic crops multiplied by seeds.

It should be pointed out that all strategies presented in this section were only tested in model plants such as tobacco and *Arabidopsis*. Therefore, no data are available on the efficiency and stability of these systems in actual crop species under agronomic conditions.

4.4 Cre-mediated excision of marker genes

In most cases, plant transformation is inefficient and transgenic cells and regenerants must be selected from a great number of non-transformed cells via incorporation of selectable marker genes. Once plant transformation is completed, these marker genes can be eliminated. There are several reasons to produce marker-free plants (Hohn et al., 2001; Hare and Chua, 2002; Miki and McHugh, 2004; Goldstein et al., 2005): marker gene removal can prevent the movement of selectable markers within the environment, simplify the regulatory process and allow the reuse of the same marker. Different methods have been identified that enable marker gene removal: co-transformation (Komari et al., 1996), transposon-dependent repositioning (Goldsbrough et al., 1993), as well as homologous (Zubko et al., 2000) and site-specific recombination (Dale and Ow, 1991). Site-specific marker gene removal will be the main topic of this section. The plant material used has been ordered according to species, supposing that this structure of the chapter might help to compare the efficiency of different methods and to choose the optimal approach for the plant to be used. Table 1 provides summarised information about Cre-site-specific marker gene elimination systems and their efficiency in different plant species.

The theoretical concept of Cre-mediated marker gene excision was proved in tobacco about twenty years ago by two research groups (Dale and Ow, 1991; Russel et al., 1992). Marker-free plants were generated by applying the Cre recombinase constitutively either via cross-pollination or a second round of transformation. The authors reported that re-transformation provided much higher recombination efficiency. This principle was also functional in the plastid genome (Corneille et al., 2001). Both methods for constitutive *cre* expression were efficient in tobacco chloroplasts, but *Agrobacterium*-mediated Cre recombinase delivery caused plastid genome rearrangements.

Transient expression vectors - *Agrobacterium*- or virus, - worked efficiently in tobacco. Simple cocultivation of transgenic tobacco leaves harbouring the marker gene with *A. tumefaciens* containing a *cre*-plasmid led to the removal of the flanked region in 0.25% of the regenerants (Gleave et al., 1999). In comparison to cocultivation technique, the agroinfiltration method greatly increased the recombination efficiency. Regenerants without marker genes were obtained with a frequency of about 34%. In 14% of plants site-specific recombination was performed without stable recombinase integration. Delivery of the Cre protein by agroinfiltration was also adopted to remove marker genes from the plastid genome (Lutz et al., 2006). Another option to perform transient *cre* expression is the use of Cre-virus vectors. The first plant Cre-virus vector was based on PVX and demonstrated high recombination rates (48-82%) in *N. benthamiana* (Kopertekh et al., 2004b). This vector was also suitable to generate marker-free tobacco plants without a regeneration step (Kopertekh et al., 2004a). The second Cre-virus vector described is based on TMV. It was functional in *N. tabacum* plants with an efficiency of about 34% (Jia et al., 2006).

Genotype	Induction factor, expression system/promoter	*cre* expression type	Excision rate	Gene of interest	Reference
Tobacco *N. tabacum*	Cross-pollination, retransformation, 35S promoter	Constitutive	ND	*luc*	Dale and Ow, 1991
Tobacco *N. tabacum*	Cross-pollination, retransformation, 35S promoter	Constitutive	95% (retransformation)	*gusA*	Russell et al., 1992
Tobacco *N. tabacum*	Cross-pollination, 35S promoter	Constitutive	19.2%	*ASAL*	Chakraborti et al., 2008
Tobacco *N. tabacum* (plastid genome)	Cross-pollination, retransformation, 35S promoter	Constitutive	ND	-	Corneille et al., 2001
Tobacco *N. benthamiana*	PVX-Cre expression vector	Transient	48-82%	*gfp*	Kopertekh et al., 2004
Tobacco *N. tabacum*	TMV-Cre expression vector	Transient	34%	*gusA*	Jia et al., 2006
Tobacco *N. tabacum*	*A. tumefaciens*-expression vector	Transient	0.25%	*gusA*	Gleave et al., 1999
Tobacco *N. tabacum* (plastid genome)	*A. tumefaciens*-expression vector	Transient	10%	*bar*	Lutz et al., 2006
Tobacco *N. benthamiana*	*A. tumefaciens*-expression vector	Transient	34%	*gfp*	Kopertekh et al., 2005
Tobacco *N. tabacum*	Heat-shock, HSP17.5E promoter from soybean	Temporal	30-80%	*gusA*	Wang et al., 2005
A. thaliana	Chemical induction, β-estradiol inducible transactivator XVE	Temporal	29-66%	*gfp*	Zuo et al., 2001
A. thaliana	Tissue-specific induction, *AP1* and *SDS* germline specific promoters	Temporal	83-100%	-	Verweire et al., 2007
Rice	Cross-pollination, 35S promoter	Constitutive	58%	*gusA*	Hoa et al., 2002
Rice	Co-cocultivation with a purified Cre-recombinase protein	Transient	26%	*gusA*	Cao et al., 2006

Rice	Heat-shock, HSP17.5E promoter from soybean	Temporal	16%		*gusA*	Khattri et al., 2011
Rice	Chemical induction, β-estradiol inducible transactivator XVE	Temporal	29.1%		*gfp*	Sreekala et al., 2005
Maize	Cross-pollination	Constitutive	ND		*cordapA*	Ow, 2007
Maize	Cross-pollination, 35S promoter Heat-shock, HSP17.5E promoter from soybean	Constitutive Temporal	ND		*gfp*	Zhang et al., 2003
Maize	Cross-pollination, Ubi promoter	Constitutive	ND		*gusA*	Kebrach et al., 2005
Wheat	Cross-pollination, 35S promoter	Constitutive	ND		-	Srivastava et al., 1999
Potato	Heat-shock, hsp70 promoter from *Drosophila melanogaster*	Temporal	4.7%		-	Cuellar et al., 2006
Potato	PVX-Cre expression vector	Transient	20-27%		*gfp*	Kopertekh et al., 2011
Brassica juncea	Cross-pollination, 35S promoter	Constitutive	ND		*gusA*	Arumugam et al., 2007
Brassica napus	Tissue-specific induction, seed-specific napin promoter from *B. napus*	Temporal	13-81%		*vstI*	Kopertekh et al., 2009
Soybean	Tissue-specific induction, *Arabidopsis app1* embryo-specific promoter	Temporal	13%		*gusA, gat*	Li et al., 2007
Tomato	Chemical induction, β-estradiol inducible transactivator XVE	Temporal	15%		*cryIAc*	Zhang et al., 2006
Tomato	Chemical induction, β-estradiol inducible transactivator XVE	Temporal	ND	*atlpk2β*		Zhang et al., 2009

ND, not determined
luc: luciferase gene
gfp: green fluorescent protein gene
atlpk2β: inositol polyphospate 6-/3-kinase gene
gus: beta-glucuronidase gene
vstI: stilbene synthase gene from *Vitis vinifera*
bar: phosphinothricin acetyltransferase gene
cryIAc: a synthetic *Bacillus thuringiensis* endotoxin gene
ASAL: allium sativum leaf agglutinin gene
gat: glyphosate acetyltransferase gene
cordapA: dihydrodipicolinate synthase gene

Table 1. Cre-based systems for marker gene elimination

Different promoters, including heat-shock and developmentally regulated ones, were tested in autoexcision vectors in tobacco. In the heat-shock inducible system, the Cre recombinase was more effective in somatic tissues in comparison to germline cells: 70-80% of the regenerants derived from heat-treated leaves lost *lox*-flanked DNA fragments, whereas only 30-40% of seeds after heat-shock gave rise to marker-free plants (Wang et al., 2005). A developmentally regulated Cre-*lox* system based on the seed-specific napin promoter was more efficient in *N. benthamiana* plants: genetic and molecular analysis of T1 progeny indicated DNA excision in all transgenic lines tested (Kopertekh et al., 2010).

Both tobacco and *Arabidopsis thaliana* served as model systems to test different gene elimination approaches. An elegant self-excision Cre-system regulated by β-estradiol was applied for the first time in *Arabidopsis* with an efficiency of 29-66% (Zuo et al., 2001). Furthermore, Verweire et al. (2007) reported an almost complete autoexcision driven by germline promoters.

Rice has been intensively studied for Cre-mediated marker gene excision. The efficiency of all three categories of methods, transient, constitutive and temporal expression, has been evaluated. In one of the first studies on the Cre-*lox* system in rice, *lox*- and *cre*-constructs were combined by cross-fertilization of transgenic plants (Hoa et al., 2002). In the Cre-*lox* hybrids from T$_2$ crosses a high marker gene deletion frequency of 58.3% was observed. Marker gene excision was also accomplished in transgenic rice cells by simple co-cultivation with a purified cell-permeable Cre recombinase protein (Cao et al., 2006). About 26% of regenerants derived from Cre-treated calli were scored as putative recombinants. However, no data are available about germinal inheritance of the recombined "footprint". Thus, it is difficult to assess the efficiency of this of this approach properly. Marker gene excision and inheritance of the excised locus were observed in one transgenic rice line containing a *lox*-target construct and a single copy of the *cre* gene under the control of the HSP17.5E heat-shock inducible promoter (Khattri et al., 2011). An obvious drawback of this co-transformation approach is the necessity to segregate the *cre*-construct after recombination. Sreekala et al. (2005) demonstrated the removal of the flanked fragment from the genome of transgenic rice in a single-step transformation by using the β–estradiol regulation of Cre. In total, 29% of transgenic T$_0$ plants were marker-free or could segregate marker-free progeny. In the Cre-*lox* system controlled by a floral specific promoter complete auto-excision was observed in three out of eight rice lines with an efficiency of 37.5% (Bai et al., 2008). This approach may be considered as the most promising for the removal of unnecessary sequences in rice since (i) Cre expression is restricted to a special tissue, (ii) recombined lines can be obtained without crossing or additional treatment and (iii) this one-step transformation approach provides high recombination frequencies.

Two strategies - cross-pollination and heat-shock inducible autoexcision - have been shown to be useful to develop transgenic maize plants harbouring only the trait gene. The crossing strategy worked with nearly 100% efficiency in several laboratories (Zahn et al., 2003; Kebrach et al., 2005). Moreover, commercial marker-free maize LYO38 was developed by Monsanto through sexual crossing between *lox*- and *cre*-plants following segregation of the *cre* gene in the next generation. A comparative study by Zhang et al. (2003) also demonstrated that autoexcision induced by heat-shock provided precise, complete and stable marker gene excision.

There is less information available on Cre-mediated marker gene elimination in wheat. Srivastava et al. (1999) combined two potential applications of site-specific recombination in one plant vector. Transgenic wheat plants harbouring a DNA fragment between mutant *lox*511 sites in opposite orientation and a marker gene between wild type *lox* sites in direct orientation were crossed with a *cre*-transgenic line. Some T_1 plants without the selection marker and a reduced copy number were detected by PCR.

The feasibility of the Cre-*lox* system for the removal of marker genes in *Brassicaceae* was demonstrated in two studies. In the first one, the *lox* sites and *cre* gene under control of a constitutive promoter were combined by cross-pollination to produce marker-free *Brassica juncea* plants (Arumugam et al., 2007). The Cre recombinase displayed low activity in meristematic cells. Thus, an additional regeneration step from leaf explants was necessary to obtain *B. juncea* plants without marker genes. The application of seed-specific napin promoter from *B. napus* to control the *cre* gene seems to be more suitable to perform the germline transmission of the recombination event (Kopertekh et al., 2009). Marker-free *B. napus* plants could be generated with high efficiency (13-81%).

Two techniques, *cre* induction by heat-shock and PVX-Cre-expression have been optimized for vegetatively propagated potato. About 4% of regenerated shoots derived from heat treated internodes and tubes demonstrated the marker-free phenotype (Cuellar et al., 2007). Transient PVX-Cre-based expression resulted in a more efficient excision of the *npt*II gene cloned between recognition sites (Kopertekh et al., 2011). Excision rates of 20-27% were achieved by applying the particle bombardment infection method and the P19 silencing suppressor protein.

In the auto-excision Cre-system developed for soybean transformation, a selectable marker gene was expressed at an early transformation step and then removed by the Cre recombinase driven by *app1* embryo-specific promoter from *A. thaliana* (Li et al., 2007). This excision reaction led to the activation of the glyphosate acetyltransferase (*gat*) gene. It was shown that 13% of events exhibited complete excision of the marker gene.

The application of a chemically-regulated autoexcision Cre-system in tomato was reported by Zhang et al. (2006). β-estradiol treatment resulted in the excision of *cre* and marker genes and subsequently in the fusion of the endotoxin gene *cryIA* with the promoter sequence. 15% of T_1 progeny plants harboured a marker-free phenotype.

Generally, the newly designed Cre-systems have first been tested in tobacco and *Arabidopsis*, and subsequently extended to actual crops. It should be pointed out that the same Cre-systems demonstrate higher recombination efficiencies in model species in comparison to agricultural crops. For example, for the β-estradiol inducible self-excision Cre system in *Arabidopsis* 29-66% efficiency was observed, whereas in rice and tomato only 30% and 15%, respectively. Similarly, heat-shock induction resulted in 30-80% excision rates in tobacco and only 4% in potato. Recently, this tendency was also demonstrated for the transient PVX-Cre vector. In comparison to *N. benthamiana* (48-82%), lower excision rates of 20-27% were shown for potato (Kopertekh et al., 2011).

5. Conclusions

Since the initial work of Dale and Ow (1990) demonstrating the functionality of the Cre-*lox* system in plant cells, a number of technologies based on site-specific recombination have been developed, tested and implemented into transformation protocols. All these technologies rely on two basic genome modifications caused by Cre recombinase:

integration or removal of foreign DNA fragments. In this review we focused on the employment of Cre-mediated elimination of transgene sequences. The literature analysis indicates several current trends in optimizing the recombination strategies and their practical application. First, future employment of the Cre-*lox* system will likely incorporate more precise temporal and spatial control of *cre* expression. To this end a number of conditional and transient excisional Cre-systems have been designed and tested during the last decade (see the paragraph "Cre-expression strategies: efficiency and limitations"). Second, the methodological progress mentioned above allowed extending the recombination technology from model (tobacco and *Arabidopsis* plants) to actual crops, including generatively and vegetatively propagated species, monocots and dicots. Third, the excisional recombination method was combined with trait genes, illustrating the development from laboratory experiments to practical utilization; this tendency can mainly be observed for the marker gene elimination technology. Among commercial traits combined with the Cre-*lox* system are modification of protein composition (Ow, 2007), tolerance to environmental stress (Zhang et al., 2009) as well as herbicide (Lutz et al., 2006; Li et al., 2007) and insect (Chakraborti et al., 2008; Zhang et al., 2006) resistance. The first marker-free commercial maize event LY038, which received the US regulatory approval in 2006, provides higher lysine content (Ow, 2007).

However, the approval process for the commercial utilization of genetically modified plants based on the techniques described above might require additional regulatory costs. The first consideration could be connected to the possible reintegration/persistence of excised DNA fragments. Although it is generally assumed that the elimination product is lost upon cell division there is one report showing the presence of deleted DNA as an extra-chromosomal circle in wheat cells (Srivastava and Ow, 2003). The second consideration is linked with possible unintended effects which might be caused by the Cre-*lox* system. Numerous reports exist that demonstrate high specificity of Cre-mediated recombination. Nevertheless, several articles have described undesirable Cre-mediated changes in mammalian genomes (Schmidt et al., 2000; Thyagarajan et al., 2000; Loonstra et al., 2001; Silver and Livingston, 2001). The impact of Cre activity on the plant genome is not well studied. Two types of effects have been described: phenotypic abnormalities and DNA rearrangements in chloroplasts. In petunia, tomato, tobacco and *Arabidopsis* aberrant phenotypes such as leaf chlorosis, growth retardation and reduced fertility were associated with high levels of *cre* expression (Coppoolse et al., 2003). These phenotypic abnormalities were not connected with chromosomal rearrangements: they always co-segregated with the *cre* transgene. In contrast, non-specific DNA recombination products have been identified in the plastid genome by two research groups (Corneille et al., 2001; Hajdukiewicz et al., 2001). Temporal or developmental regulation of the Cre activity would decrease or eliminate these side-effects and subsequently simplify risk assessment process. Another concern related to the Cre-*lox* application is the presence of a *lox* recognition site in the final product. Theoretically, non-predicted recombination between this *lox* site and pseudo-*lox* sites in the genome can occur in the presence of the Cre protein. In fact, the probability of such an event is extremely low. First, the recombination reaction strongly depends on a sequence similarity between the introduced *lox* and genomic pseudo-*lox* sites. The recombination efficiency is greatly reduced when only a few nucleotides in the *lox* spacer region are different (Hoess et al., 1986). Second, the distance between recombination sites plays an important role: the recombination between *lox* sites located at unlinked chromosomes is less efficient (Qin et al., 1994).

Despite the regulatory issues described above, we expect that site-specific excisional recombination will become a routine method in plant biotechnology and find a broader application for the commercial use of crop plants.

6. References

Arumugam N, Gupta V, Jagannath A, Mukhopadhyay A, Pradhan AK, Burma PK and Pental D (2007) A passage through in vitro culture leads to efficient production of marker-free transgenic plants in Brassica juncea using the Cre/loxP system. Trangenic Res. 16(6): 703-712, DOI: 10.1007/s11248-006-9058-7

Austin S, Ziese M and Sternberg N (1981) A novel role of site-specific recombination in maintenance of bacterial replicons. Cell 25(3): 729-736, DOI:10.1016/0092-8674(81)90180-X

Bai X, Wang Q and Chu C (2008) Excision of a selective marker in transgenic rice using a novel Cre-lox system controlled by a floral specific promoter. Transgenic Res. 17(6): 1035-1043, DOI: 10.1007/s11248-008-9182-7

Baubonis W and Sauer B (1993) Genomic targeting with purified Cre recombinase. Nucleic Acids Research 21(9): 2025-2029, DOI: 10.1093/nar/21.9.2025

Bayer M and Hess D. (2005) Restoring full pollen fertility in transgenic male-sterile tobacco (Nicotiana tabacum L.) by Cre-mediated site-specific recombination. Mol. Breeding 15(2): 193–203, DOI: 10.1007/s11032-004-5042-1

Bayley CC, Morgan M, Dale E and Ow DW (1992) Exchange of gene activity in transgenic plants catalyzed by the Cre-lox site-specific recombination system. Plant Mol Biol. 18(2): 353-361, DOI: 10.1007/BF00034962

Cao MX, Huang JQ, Yao QH, Liu SJ, Wang CL and Wei ZM (2006) Site-Specific DNA Excision in Transgenic Rice With a Cell-Permeable Cre Recombinase. Molecular Biotechnology 32(1): 55–64, DOI: 10.1385/MB:32:1:055

Chakraborti D, Sarkar A, Mondal HA, Schuermann D, Hohn B, Bidyut K. Sarmah BK, Das S (2008) Cre/lox system to develop selectable marker free transgenic tobacco plants conferring resistance against sap sucking homopteran insect. Plant Cell Rep 27(10): 1623–1633, DOI 10.1007/s00299-008-0585-y

Coppoolse ER, de Vroomen MJ, Roelofs D, Smit J, van Gennip F, Hersmus BJM, Nijkamp HJ and van Haaren MJJ (2003) Cre recombinase expression can result in phenotypic aberrations in plants. Plant Mol Biol. 51(2): 263-279, DOI: 10.1023/A:1021174726070

Corneille S, Lutz K, Svab Z and Maliga P (2001) Efficient elimination of selectable marker genes from the plastid genome by the Cre-lox site-specific recombination system. Plant J. 27(2): 171-178, DOI: 10.1046/j.1365-313x.2001.01068.x

Cuellar W, Gaudin A, Solorzano D, Casas A, Nopo L, Chudalayandi P, Medrano G, Kreuze J and Ghislain M (2006) Self-excision of the antibiotic resistance gene nptII using a heat inducible Cre-loxP system from transgenic potato. Plant Mol Biol. 62(1-2): 71-82, DOI: 10.1007/s11103-006-9004-3

Dale EC, Ow DW (1991) Gene transfer with subsequent removal of the selection gene from the host genome. Proc Natl Acad Sci USA 88(23): 10558–10562. DOI:10.1073/pnas.88.23.10558

Dale, EC and Ow, DW (1990) Intra- and intermolecular sitespecific recombination in plant cells mediated by bacteriophage P1 recombinase. Gene 91(1): 79–85, DOI 10.1016/0378-1119(90)90165-N

Daniell H (2002) Molecular strategies for gene containment in transgenic crops. Nat Biotechnol. 20: 581-586, DOI:10.1038/nbt0602-581

De Block M, Debrouwer D, Moens T (1997) The development of a nuclear male sterility system in wheat. Expression of the barnase gene under the control of tapetum specific promoters. Theor Appl Genet. 95(1-2): 125-131, DOI: 10.1007/s001220050540

De Buck S and Depicker A (2004) Gene expression and level of expression. In P Christou, H Klee, eds, Handbook of Plant Biotechnology, Vol 1. John Wiley & Sons, Chichester, UK, pp 331-345

De Buck S, Peck I, De Wilde C, Marjanac, G, Nolf J, De Paepe A and Depicker, A (2007) Generation of single-copy T-DNA transformants in Arabidopsis by the CRE/loxP recombination-mediated resolution system. Plant Physiol. 145(4): 1171-1182, DOI: 10.1104/pp.107.104067

De Buck S, Van Montagu M and Depicker A. (2001) Transgene silencing of invertedly repeated transgenes is released upon deletion of one of the transgenes involved. Plant Mol. Biol. 46(4): 433-445, DOI: 10.1023/A:1010614522706

De Paepe A, De Buck S, Hoorelbeke K, Nolf J, Peck I and Depicker A (2009) High frequency of single-copy T-DNA transformants produced by floral dip in CRE-expressing Arabidopsis plants. Plant J. 59(4): 517-527 DOI: 10.1111/j.1365-313X.2009.03889.x

Fladung M, Becker D (2010) Targeted integration and removal of transgenes in hybrid aspen (Populus tremula L. 9 P. tremuloides Michx.) using site-specific recombination systems. Plant Biol. 12(2): 334-340, DOI: 10.1111/j.1438-8677.2009.00293.x

Gidoni D, Srivastava, V and Carmi, N (2008) Site-specific excisional recombination strategies for elimination of undesirable transgenes from crop plants. In Vitro Cellular & Developmental Biology - Plant, 44(6): 457-467, DOI: 10.1007/s11627-008-9140-3

Gilbertson, L (2003) Cre-lox recombination: Cre-active tools for plant biotechnology. Trends Biotechnol. 21(12): 550-555, DOI:10.1016/j.tibtech.2003.09.011

Gleave AP, Mitra DS, Mudge SR and Morris BAM (1999) Selectable marker-free transgenic plants without sexual crossing: transient expression of cre recombinase and use of a conditional lethal dominant gene. Plant Mol Biol. 40(2): 223-235, DOI: 10.1023/A:1006184221051

Goldsbrough AP, Lastrella CN and Yoder JI (1993) Transposition-mediated re-positioning and subsequent elimination of marker genes from transgenic tomato. Nat Biotechnology 11: 1286-1292, DOI:10.1038/nbt1193-1286

Goldstein, DA, Tinland B, Gilbertson LA, Staub, JM, Bannon GA, Goodman, RE, McCoy RL and Silvanovich A (2005) Human safety and genetically modified plants: a review of antibiotic resistance markers and future transformationselection technologies. J Applied Microbiology 99(1): 7-23, DOI: 10.1111/j.1365-2672.2005.02595.x

Grindley NDF, Whiteson KL and Rice PA (2006) Mechanisms of Site-Specific Recombination. Annual Review of Biochemistry 75: 567-605, DOI: 10.1146/annurev.biochem.73.011303.073908

Guo F, Gopaul DN and van Duyne G D (1999) Asymmetric DNA bending in the Cre-loxP site-specific recombination synapse. Proc Natl Acad Sci USA 96(13): 7143-7148, DOI: 10.1073/pnas.96.13.7143

Guo F, Gopaul, GN and van Duyne, GD (1997) Structure of Cre recombinase complexed with DNA in a site-specific recombination synapse. Nature. 389(6646): 40-46; DOI: 10.1038/37925

Hajdukiewicz PTJ, Gilbertson L. and Staub JM (2001) Multiple pathways for Cre/lox-mediated recombination in plastids. Plant J. 27(2): 161-170, DOI: 10.1046/j.1365-313x.2001.01067.x

Hansen G, Chilton M-D (1996) "Agrolistic" transformation of plant cells: integration of T-strands generated in planta. Proc Natl Acad Sci USA 93(25): 14978–14983

Hare PD and Chua N-H (2002) Excision of selectable marker genes from transgenic plants. Nature Biotechnol. 20: 575-580, DOI:10.1038/nbt0602-575

Hoa TTC, Bong BB, Huq E and Hodges TK (2002) Cre-lox site-specific recombination controls the excision of a transgene from the rice genome. Theor Appl Genet. 104(4): 518-525, DOI: 10.1007/s001220100748

Hochman L, Segev N, Sternberg N and Cohen G. (1983) Site-Specific Recombinational Circularization of Bacteriophage Pl DNA. Virology 131(1): 11-17, DOI: 10.1016/0042-6822(83)90528-7

Hoess RH and Abremski K (1984) Interaction of the bacteriophage P1 recombinase Cre with the recombining site loxP. Proc Natl Acad Sci USA 81(4): 1026-1029

Hoess RH, Wierzbicki A, and Abremski K (1986) The role of the loxP spacer region in P1 site-specific recombination. Nucleic Acids Res. 14(5): 2287-2300, DOI: 10.1093/nar/14.5.2287

Hoess RH, Ziese M, Sternberg N. (1982) P1 site-specific recombination: nucleotide sequence of the recombining sites. Proc Natl Acad Sci USA. 79(11): 3398-3402

Hoff T, Schnorr KM and Mundy J (2001) A recombinase-mediated transcriptional induction system in transgenic plants. Plant Mol Biol. 45(1): 41-49, DOI: 10.1023/A:1006402308365

Hohn B, Levy AA and Puchta H (2001) Elimination of selection markers from transgenic plants. Curr. Opin. Biotechnol. 12(2): 139-143, DOI:10.1016/S0958-1669(00)00188-9

Jia H, Pang Y, Chen X and Fang R (2006) Removal of the selectable marker gene from transgenic tobacco plants by expression of cre recombinase from a tobacco mosaic virus vector through agroinfection. Transgenic Res. 15(3): 375-384, DOI: 10.1007/s11248-006-0011-6

Joubès J, De Schutter K, Verkest A, Inzé D and De Veylder L (2004) Conditional, recombinase-mediated expression of genes in plant cell cultures. Plant J. 37(6): 889–896, DOI: 10.1111/j.1365-313X.2004.02004.x

Kebrach S, Lörz H and Becker D (2005) Site-specific recombination in Zea mays. Theor Appl Genet. 111(8): 1608-1616, DOI: 10.1007/s00122-005-0092-2

Keenan RJ and Stemmer WPC (2002) Nontransgenic crops from transgenic plants. Nat Biotechnol. 20: 215-216, DOI:10.1038/nbt0302-215

Khattri A, Nandy S, Srivastava V (2011) Heat-inducible Cre-lox system for marker excision in transgenic rice. J Biosci. 36(1): 37-42, DOI 10.1007/s12038-011-9010-8

Kilby NJ, Davies GJ, Snaith MR and Murray JAH (1995) FLP recombinase in transgenic plants: constitutive activity in stably transformed tobacco and generation of marked cell clones in Arabidopsis. Plant J. 8(5): 637–652, DOI: 10.1046/j.1365-313X.1995.08050637.x

Kittiwongwattana C, Lutz K, Clark M, Maliga P (2007) Plastid marker gene excision by the phiC31 phage site-specific recombinase. Plant Mol Biol. 64(1-2): 137-143, DOI: 10.1007/s11103-007-9140-4

Komari T, Hiei Y, Saito Y, Murai N and Kumashiro T (1996) Vectors carrying two separate T-DNAs for co-transformation of higher plants mediated by Agrobacterium tumefaciens and segregation of transformants free from selection markers. Plant J. 10(1): 165-174, DOI: 10.1046/j.1365-313X.1996.10010165.x

Kopertekh L, Broer I and Schiemann J (2009) Developmentally regulated site-specific marker gene excision in transgenic B. napus plants. Plant Cell Rep. 288(7): 1075–1083, DOI: 10.1007/s00299-009-0711-5

Kopertekh L, Jüttner G and Schiemann J (2004a) Site-specific recombination induced in transgenic plants by PVX virus vector expressing bacteriophage P1 recombinase. Plant Science 166(2): 485-492, DOI:10.1016/j.plantsci.2003.10.018

Kopertekh L, Jüttner G and Schiemann J (2004b) PVX-Cre-mediated marker gene elimination from transgenic plants. Plant Mol Biol 55(4): 491-500, DOI: 10.1007/s11103-004-0237-8

Kopertekh L, Schulze K, Frolov A, Strack D, Broer I, Schiemann J. (2010) Cre-mediated seed-specific transgene excision in tobacco. Plant Mol Biol. 72(6): 597-605, DOI: 10.1007/s11103-009-9595-6

Kopertekh L, v. Saint Paul V, Krebs E and Schiemann J (2011) Utilization of PVX-Cre expression vector in potato. Transgenic Res., submitted for publication

Kopertekh L. and Schiemann J. (2005) Agroinfiltration as a tool for transient expression of cre recombinase in vivo. Transgenic Res. 14(5): 793-798, DOI: 10.1007/S11248-011-9558-y

Li Z, Xing A, Moon BP, Burgoyne SA, Guida AD, Liang H, Lee C, Caster CS, Barton JE, Klein TM, Falko SC (2007) A Cre/ loxP -mediated self-activating gene excision system to produce marker gene free transgenic soybean plants. Plant Mol Biol. 65(3): 329-341, DOI: 10.1007/s11103-007-9223-2

Lloyd AM and Davis RW (1994) Functional expression of the yeast FLP/FRT site-specific recombination system in Nicotiana tabacum. Mol. Gen. Genet. 242(6): 653–657, DOI: 10.1007/BF00283419

Loonstra A, Vooijs M, Beverloo HB, Allak BA, van Drunen E, Kanaar R, Berns A and Jonkers J. (2001) Growth inhibition and DNA damage induced by Cre recombinase in mammalian cells. Proc. Natl. Acad. Sci. USA 98(16): 9209-9214, DOI: 10.1073/pnas.161269798

Luo K, Duan H, Zhao D, Zheng X, Deng W, Chen, Y, Stewart C N, McAvoy R, Jiang X, Wu Y, He A, Pei Y. and Li Y. (2007) "GM-gene-deletor": fused loxP-FRT recognition sequences dramatically improve the efficiency of FLP or Cre recombinase on transgene excision from pollen and seed of tobacco plants. Plant Biotechnol J. 5(2): 263-274, DOI: 10.1111/j.1467-7652.2006.00237.x

Lutz KA, Bosacchi MH, Maliga P (2006) Plastid marker-gene excision by transiently expressed CRE recombinase. Plant J. 45(3): 447–456, DOI: 10.1111/j.1365-313X.2005.02608.x

Lyznick LA, Mitchell JC, Hirayama L and Hodges TK (1993) Activity of yeast FLP recombinase in maize and rice protoplasts. Nucleic Acids Res. 21(4): 969-975, DOI: 10.1093/nar/21.4.969

Lyznik LA, Gordon-Kamm WJ and Tao Y (2003) Site-specific recombination for genetic engineering in plants. Plant Cell Reports 21(10): 925-932, DOI: 10.1007/s00299-003-0616-7

Maeser S and Kahmann R (1991) The Gin recombinase of phage Mu can catalyze site-specific recombination in plant protoplasts. Mol. Gen. Genet. 230(1-2): 170–176, DOI: 10.1007/BF00290665

Miki B and McHugh S (2004) Selectable marker genes in transgenic plants: applications, alternatives and biosafety. J Biotechnology 107(3): 193-232, DOI:10.1016/j.jbiotec.2003.10.011

Mlynarova L, Conner AJ and Nap J-P. (2006) Directed microspore-specific recombination of transgenic alleles to prevent pollen-mediated transmission of transgenes. Plant Biotech J. 4(4): 445-452, DOI: 10.1111/j.1467-7652.2006.00194.x

Moon HS, Abercrombie LL, Eda S, Blanvillain R, Thomson JG, Ow DW, Stewart CN Jr (2011) Transgene excision in pollen using a codon optimized serine resolvase CinH-RS2 site-specific recombination system. Plant Mol Biol. 75(6): 621-631, DOI: 10.1007/s11103-011-9756-2

Moon HS, Li Y and Stewart Jr CN (2009) Keeping the genie in the bottle: transgene biocontainment by excision in pollen. Trends in Biotechnol. 28(1): 3-8, DOI:10.1016/j.tibtech.2009.09.008

Moravčíková J, Vaculková E, Bauer M and Libantová J (2008) Feasibility of the seed specific cruciferin C promoter in the self excision Cre/loxP strategy focused on generation of marker-free transgenic plants. Theor Appl Genet. 117(8): 1325-1334, DOI: 10.1007/s00122-008-0866-4

Odell J, Caimi P, Sauer B and Russell, S. (1990) Site-directed recombination in the genome of transgenic tobacco. Mol Gen Genet 223(3): 369-378, DOI: 10.1007/BF00264442

Odell J, Hoopes JL and Vermerris W (1994) Seed-specific gene activation mediated by the Cre/lox site-specific recombination system. Plant Physiol. 106(2): 447-458, DOI: 10.1104/pp.106.2.447

Onouchi H, Yokoi K, Machida C, Matzuzaki H, Oshima Y, Matsuoka K, Nakamura K and Machida Y (1991) Operation of an efficient site-specific recombination system of Zygosaccharomyces rouxii in tobacco cells. Nucl. Acids Res. 19(23): 6373–6378, DOI: 10.1093/nar/19.23.6373

Ow DW (2002) Recombinase-directed plant transformation for the post-genomic era. Plant Mol Biol. 48(1-2): 183-200, DOI: 10.1023/A:1013718106742

Ow, DW (2007) GM maize from site-specific recombination technology, what next? Current Opinion in Biotechnology 18(2): 115-120, DOI:10.1016/j.copbio.2007.02.004

Qin M, Bayley C, T Stockton T and Ow DW (1994) Cre recombinase-mediated site-specific recombination between plant chromosomes. Proc. Natl. Acad. Sci. USA 91(5): 1706-1710, DOI: 10.1073/pnas.91.5.1706

Rubtsova M, Kempe, K, Gils, A, Ismagul, A, Weyen J and Gils, M (2008) Expression of active Streptomyces phage phiC31 integrase in transgenic wheat plants. Plant Cell Rep. 27(12): 1821-1831, DOI: 10.1007/s00299-008-0604-z

Russell SH, Hoopes JL and Odell JT (1992) Directed excision of a transgene from the plant genome. Mol. Genet. Genet. 234(1): 49-59, DOI: 10.1007/BF00272344

Schmidt EE, Taylor DS, Prigge JR, Barnett S and Capecchi MR (2000) Illegitimate Cre-dependent chromosome rearrangements in transgenic mouse spermatids. Proc. Natl. Acad. Sci. USA 97(25): 13702-13707, DOI: 10.1073/pnas.240471297

Segev N and Cohen N (1981) Control of Circularization of Bacteriophage Pl DNA in Escherichia coli. Virology 114(2): 333-342, DOI:10.1016/0042-6822(81)90215-4

Silver DP and Livingston DM (2001) Self-excising retroviral vectors encoding the Cre recombinase overcome Cre-mediated cellular toxicity. Mol. Cell. 8(1): 233-243, DOI:10.1016/S1097-2765(01)00295-7

Sreekala C, Wu L, Gu Wang D and Tian D (2005) Excision of a selectable marker in transgenic rice (*Oryza sativa* L.) using a chemically regulated Cre-*loxP* system. Plant Cell Rep. 24(2): 86-94, DOI: 10.1007/s00299-004-0909-5

Srivastava V and Ow DW (2001) Single-copy primary transformants of maize obtained through the co-introduction of a recombinase-expressing construct. Plant Mol Biol 46(5): 561–566, DOI: 10.1023/A:1010646100261

Srivastava V and Ow DW (2003) Rare instances of Cre-mediated deletion product maintained in transgenic wheat. Plant Mol Biol 52(3): 661-668, DOI: 10.1023/A:1024839617490

Srivastava V, Anderson OD, Ow DW (1999) Single-copy transgenic wheat generated through the resolution of complex integration patterns. Proc Natl Acad Sci USA 96(20): 11117–11121, DOI: 10.1073/pnas.96.20.11117

Srivastava V, Vasil V and Vasil IK (1996) Molecular characterization of the fate of transgenes in transformed wheat. Theor. Appl. Genet. 92(8): 1031-1037, DOI: 10.1007/BF00224045

Thomson JG, Yau YY, Blanvillain R, Chiniquy D, Thilmony R and Ow DW (2009) ParA resolvase catalyzes site-specific excision of DNA from the Arabidopsis genome. Transgenic Res. 18(2): 237–248, DOI 10.1007/s11248-008-9213-4

Thyagarajan B, Guimaraes MJ, Groth AC and Calos MP (2000) Mammalian genomes contain active recombinase recognition sites. Gene 244(1-2): 47-54, DOI:10.1016/S0378-1119(00)00008-1

Tremblay A, Beauchemin C, Seguin A and Laliberte J-F. (2007) Reactivation of an integrated disabled viral vector using a Cre-loxP recombination system in Arabidopsis thaliana. Transgenic Res. 16(2): 213–222, DOI: 10.1007/s11248-006-9038-y

Van Ex F, Verweire D, Claeys M, Depicker A, Angenon G (2009) Evaluation of seven promoters to achieve germline directed Cre-lox recombination in Arabidopsis thaliana. Plant Cell Rep. 28(10): 1509-1520, DOI: 10.1007/s00299-009-0750-y

Vergunst AC, Schrammeijer B, den Dulk-Ras A, De Vlaam CMT, Regensburg-Tuink TJG and Hooykaas PJJ (2000) Vir B/D4 – dependent protein translocation from Agrobacterium into plant cells. Science 290(5493): 979-982, DOI: 10.1126/science.290.5493.979

Verweire D, Verleyen K, De Buck S, Claeys M and Angenon G (2007) Marker-free transgenic plants through genetically programmed auto-excision. Plant Physiol. 145(4): 1220-1231, DOI: 10.1104/pp.107.106526

Wang M-B and Waterhouse PM (2000) High-efficiency silencing of a β-glucuronidase gene in rice is correlated with repetitive transgene structure but is independent of DNA methylation. Plant Mol. Biol. 43(1): 67–82, DOI: 10.1023/A:1006490331303

Wang Y, Chen B, Hu Y, Li J and Lin Z (2005) Inducible excision of selectable marker gene from transgenic plants by the Cre-lox site-specific recombination system. Transgenic Res. 14(5): 605-614, DOI: 10.1007/s11248-005-0884-9

Wang Y, Yau YY, Perkins-Balding D, Thomson JG (2011) Recombinase technology: applications and possibilities. Plant Cell Rep. 30(3): 267-285, DOI: 10.1007/s00299-010-0938-1

Zhang W, Subbarao S, Addae P, Shen A, Armstrong C, Peshke V and Gilbertson L. (2003). Cre-lox-mediated marker gene excision in transgenic maize (Zea mays L.) plants. Theor Appl Genet. 107(7): 1157-1168, DOI: 10.1007/s00122-003-1368-z

Zhang Y, Li H, Ouyang B, Lu Y and Ye Z (2006) Chemical-induced autoexcision of selectable markers in elite tomato plants transformed with a gene conferring resistance to lepidopteran insects. Biotechnol Lett. 28(16): 1247-1253, DOI: 10.1007/s10529-006-9081-z

Zhang Y, Liu H, Li B, Zhang JT, Li Y and Zhang H (2009) Generation of selectable marker-free transgenic tomato resistant to drought, cold and oxidative stress using the Cre/loxP DNA excision system. Transgenic Res 18(4): 607–619, DOI 10.1007/s11248-009-9251-6

Zubko E, Scutt C and Meyer P (2000) Intrachromosomal recombination between attP regions as a tool to remove selectable marker genes from tobacco transgenes. Nat. Biotechnol. 18: 442-445, DOI:10.1038/74515

Zuo J, Niu Q-W, Moller SG and Chua N-H (2001) Chemical-regulated, site-specific DNA excision in transgenic plants. Nat. Biotechnol. 19: 157-161, DOI:10.1038/84428

11

GMO Safety Assessment-Feasibility of Bioassay to Detect Allelopathy Using Handy Sandwich Method in Transgenic Plants

Katsuaki Ishii[1], Akiyoshi Kawaoka[2] and Toru Taniguchi[1]
[1]Forest Bio-Research Center, Forestry and Forest Products Research Institue
[2]Nippon Paper Institute, Pulp and Paper Research Laboratory
Japan

1. Introduction

Recently, there are many reports on production and field test of transgenic plants including forest trees (Hinchee *et al.*, 2011, Kole and Hall, 2008, Walter *et al.*, 2004). However there are arguments and delicate matters about field release of transgenic trees concerning the influence to environment (McLean and Charrest, 2000). Regulations on recombinant DNA plant biotechnology were developed in USA, Canada, Europe, Oceania, China and Japan (Strauss, 2003, Kalaitzandonakes, 2004, Redenbaugh and McHughen, 2004, Lu & Hu, 2011, Watanabe et al., 2004) as well as in international agreement (Strauss *et al.*, 2009). The Conference of the Parties 9 meeting held in Milan 2003 decided recognizing host parties evaluate risks associated with the use of genetically modified organisms by afforestaion and reforestation project activities. Risk assessments shall be carried out in a scientifically sound manner and taking into account recognized risk assessment techniques to identify and evaluate the possible adverse effects of living modified organisms on the conservation and sustainable use of biological diversity (Strauss *et al.*, 2009). Environmental safety is considered in speed of degradation of introduced genes in the soil, impact on soil invertebrates like earthworms, impact on aquatic invertebrates like daphnia, impact on beneficial insects like ladybugs, impact on fish, birds, mammals and other plants. Eucalyptus species has in general allelopathy activity like suppression of growth and germination of understory weeds (Zeng *et al.*, 2008). Poplar has also some allelopathy to the crops like wheat or mycorrhizal fungi in the field (Singh et al., 1993, Olsen *et al.*, 1971).

Transgenic trees should be introduced into commerce after they have been critically evaluated for environmental safety. Allelopathy is one of the environmental effects of plants to other plants. It is important to elucidate the effects of the transformations on the allelopathy activity. Here we report the application of a handy sandwich-type bioassay method which was efficient with crop species (Fujii et al., 2003, Golisz et al., 2007) to assess the allelopathy of transgenic aspen trees containing anti-sense peroxidase (*prx*) (Yahong *et al.*, 2001) and antibiotic (neomycin phosphotransferase ; *nptII*) (Bevan *et al.*, 1983) gene, and eucalyptus trees containing transcriptional element (*Ntlim*1) (Kawaoka *et al.*, 2000, 2006) and antibiotic (*nptII*) gene. The purpose of this study is to offer possible screening method to eliminate high allelopathy transformants before field plantation.

This is the first report on application of the new bioassay of handy sandwich method to assess the allelopathy of transgenic aspen and eucalyptus.

2. Materials and methods

2.1 Transgenic trees and control

Transgenic aspen (*Populus sieboldii* x *grandidentata*)containing anti-sense *prx* and antibiotic (*nptII*) gene, and eucalyptus (*Eucalyptus camaldulensis*) (Fig. 1) containing transcriptional element (*Ntlim*1) and antibiotic (*nptII*) gene were used as experimental targets. Both transgenic approaches were intended to create less lignin wood for efficient pulping. Control tree of aspen was the same clone Y-63 which was transformed, and that of eucalyptus was seedling of same species.

Fig. 1. Transgenic *Eucalyptus camaldulensis*.

2.2 Confirmation of transcription of introduced genes

RNA was extracted from leaves and stems of individual transformants with a QIAGEN RNeasy Plant Mini Kit (QIAGEN, Hilden, Germany) according to the protocol provided by the manufacturer using lysis buffers which contains guanidine thiocyanide and RNA adsorption membrane. One microgram of total RNA was reverse transcribed and then the cDNA amplified with a QIAGEN One Step RT-PCR Kit (QIAGEN, Hilden, Germany) according to the protocol provided by the manufacturer. The oligonucleotide primers used for RT-PCR were 5'- AAACAATTACCAACACTACC-3' (forward) and 5'- ACCTGAAAGGGCAACCAGGT-3' (reverse) with anti-peroxidase gene, and 5'-

GAGGCTATTCGGCTATGACT-3′ (forward) and 5′-AATCTCGTGATGGCAGGTTG-3′ (reverse) with *npt II* gene. Conditions for amplification were reverse transcription at 50 ºC for 30 min, initial PCR activation step at 95 ºC for 15 min followed by 35 cycles at 94 ºC for 1 min, 60 ºC for 1 min, and 72 ºC for 1 min. The amplified fragments of cDNA was subjected to electrophoresis in a 2 % (w/v) agarose gel and detected with ethidium bromide.

2.3 Allelopathy level determination by sandwich method

The sandwich method assays the allelopathic activity of leaches from dried leaves on seed germination and growth of receptor plants such as lettuce (Fujii et al., 2003, 2004). Sandwich method was done according to that of Fujii et al. (2003) and Golisz et al. (2007). They used this bioassay method to screen large number plants to detect allelopathy in the 239 medicinal plant species. They demonstrated this method is a less time-consuming bioassay method and could be used to screen a large number of samples. In our experiment, one year old leaves of the transgenic and control trees of *E. camaludulensis* and *P. sieboldii x grandidentata* grown in the containment green house were collected and oven-dried at 60 ºC for over night, and then 50 mg in dry weight of them sandwiched between the layers of 0.5 % low melting point (31 ºC) agar in the multi-well dishes (6 wells whose diameter is 35 mm). Leaves of 1 year old *E. cinerea* was also used as reference in the experiment of aspen. Six replicated samples were used for each transformant and control.

Seeds of the lettuce of Great Lakes 366 were sown on the agar bed, and then germinated under constant temperature at 20 ºC. Five seeds were sown in each well.After 60 hours incubation in the dark, the length of the hypocotyls and roots was measured respectively. The inhibition rate by comparing that of control (water) was considered as allelopathy level of each sample, i.e. less growth rate indicates the high allelopathy level. Buckwheat might be also used as an indicator plant of allelopathy like lettuce (Fig. 2).

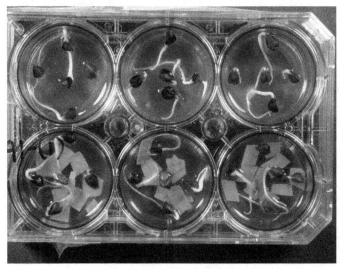

Fig. 2. Six-well multidish plastic plate used for sandwich method.

2.4 Statistical analysis

Analysis of variance (ANOVA) with allelopathy (indicated as growth rate of bioassay plant) was carried out by Fisher's test. Average values were compared using a two-sided t-test.

3. Results and discussion

It was checked by RT-PCR that transferred antibiotic genes were still expressed in the leaves and stems of transgenic trees. However, anti-sense gene of peroxidase was only expressed in the stems of transgenic trees. These results are explained that a promoter for antibiotic gene was 35S of cauliflower mosaic virus and that for peroxidase was a promoter from aspen peroxidase relating to xylogenesis.

There are slight variations of allelopathy level indicated by this bio-assay among different lines of transgenic trees. In the transgenic aspen, sandwich bio-assay evaluated by lettuce root length inhibition indicated that allelopathy of transgenic materials varied from 22 to 37 % while control was 29 % (Fig. 3). These may be of some effects of different production level of metabolic substances in the similar way like other transgenic aspen where the concentrations of total flavonoids, quercetin, kaempferol and myricetin derivatives in the leaves were different between control and transgenic trees (Haggman *et al.*, 2003).However, we did not detect statistically significant differences among transgenic and control trees by analysis of variance (P > 0.12). In contrast, there was a statistically significant difference between species *Populus* and *Eucalyptus* with allelopathy (P < 0.002) which indicated the feasibility of the sandwich method to detect the different allelopathy level between species. *E. cinerea* had stronger allelopathy than *Populus* hybrid (Fig.3). There was similar tendency in assessing allelopathy level of transgenic and non-transgenic aspen using hypocotyl growth as an indicator (Fig.4). When cellulose rich transgenic white poplar (*Populus alba*) introduced with bacterial xyloglucanase gene was compared with non-transgenic control, sandwich method revealed no statistical difference between them (J-BCH, 2007).

In the case of transgenic eucalyptus, allelopathy levels of transgenic materials were from 17 to 66 % while controls were from 17 to 26 % (Figs. 5).Among transgenic lines, (2)-4 was the lowest allelopathy tree which indicated the statistically significant difference from other lines and controls (P < 0.01). However, part of the variation in allelopathy level may be not only from transformation but also from different genetic background caused by seedling materials.Average allelopathy level of eucalyptus was higher than that of aspen as 28 % versus 31 % indicated as average growth rate to water culture. In the case of salt-tolerant transgenic *Eucalyptus camaldulensis*, there was no substantial variation between nontransgenic and transgenic lines with respect to the hypocotyl growth and root elongation of lettuce in sandwich method (Kikuchi *et al.*, 2006, 2009).They also conducted the gas chromatographic and high-performance liquid chromatographic analysis to show no qualitative and quantitative difference between the transgenic and nongenetically modified genotypes of Eucalyptus. Bioassay method is superior as the primary assessment method for allellopathy in considering the simplicity, speed, low cost, and reproducibility to chemical instrumental methods.Sandwich method for allelopathy assessment might be useful as one of the criteria of environmental effects of any transgenic platns.However, drying many samples at the same time is time consuming. So, for evaluating multiple leaves, modified sandwich method used homogenized fleshy leaf samples instead of drying them (Shimazaki et al., 2009). Chemical changes by modification or processing of some compounds may occur during the process of homogenization which leads enhancement of the effect of the transgenic and non-transgenic plant on the growth of germinating seeds. Even if the chemical changes were induced during sample preparation (drying or homogenization process), it would be thought to reflect the chemical differences between the transgenic and non-transgenic plants (Shimazaki, 2009). In the floricultural plants, the transgenic carnation (*Dianthus caryophyllus*) expressing flavonoid 3', 5'-hydroxylase gene (Moon series, www.florigene.com), assessing the allelopathic substances was done by lettuce seed germination bioassay tests in soil containing carnation debris (Kikuchi, 2008).For further study, toxicology and allergenicity testing of products from introduced genes should be carried out before commercialization of transgenic plants.

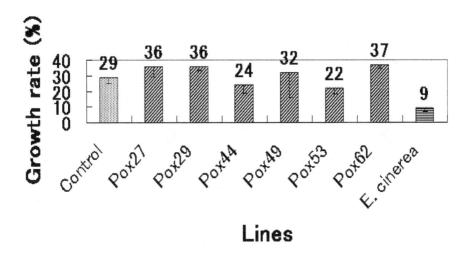

Pox 27, 29, 44, 49, 53 and 62; transgenic aspen
E; *Eucalyptus cinerea*
Bar; standard error

Fig. 3. Allelopathy level of transgenic *Populus sieboldii x grandidentata* (as % of lettuce root growth to water medium)

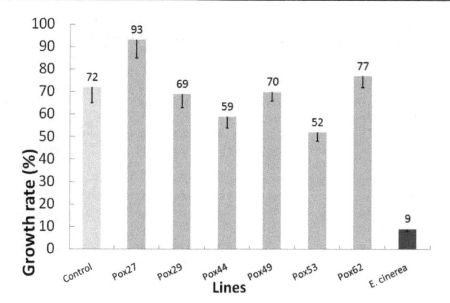

Pox 27, 29, 44, 49, 53 and 62; transgenic aspen
E; *Eucalyptus cinerea*
Bar; standard error

Fig. 4. Allelopathy level of transgenic Populus sieboldii x grandidentata (as % of lettuce hypocotyl growth to water medium)

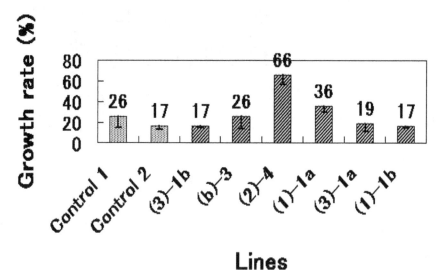

Fig. 5. Allelopathy level of transgenic *Eucalyptus camaludulensis* (as % of lettuce root growth to water medium)(3)-1b, (b)-3, (2)-4, (1)-1a, (3)-1a and (1)-1b; transgenic eucalyptus Bar; standard error

Trees have many characteristics that make them more difficult to assess them than those of agricultural crops: they have long life cycle, the production cycle may be 10 to 70 years, pollen moves over enormous distance, there are tremendous genetic and phenotypic variation and ecological complexity (McLean and Charrest, 2000). In the handling of transgenic plants, the key words are "familiarity" and "substantial equivalence". Familiarity is the knowledge of the characteristics of a plant species and the experience with the use of that plant species. Substantial equivalence is that of a novel trait within a particular plant species, in terms of its specific use and safety to the environment and human health, to those in that same species, that are in use and generally considered as safe based on valid scientific rationale. Checking of the allelopathy of transgenic trees is substantial equivalence matter. In the substantial equivalence, we must consider altered weediness potential, gene flow to related species, altered plant pest potential, potential impact on non-target organisms, and potential impact on biodiversity. Allelopathy of transgenic trees may have influence to other plants and biodiversity. The sandwich method can easily detect the high allelopathetic materials. We think, therefore, that it is useful to use this handy sandwich method as one of the criteria for assessing biosafety of plants including forest trees.

4. Conclusion

The Cartagena Protocol on Biosafety promulgated guidelines for evaluating the biosafety of living modified organisms (http://www.biodiv.org/biosafety).Transgenic plants should not be planted in the field without an environmental biosafety assessment. Commercial usage and environmental release can be permitted after evaluation in the containment growth room and field. The public concerns of environmental biosafety assessments are to define property of the transgenic plants and assess the influence of the plant on other organisms. Allelopathic influence of transgenic plants must be checked before field release.

Among the methods used for evaluating the allelopathic effects of plants are the dish pack,plant box, sandwich and soil mix methods (Shiomi et al.,1992; Yamaguchi et al., 1994; Sekine et al., 2007). Sandwich method might be one of the most handy and reliable methods for checking the transgenic plants.

5. Acknowledgment

We thank Dr. K. Fujii, National Institute for Agro-Environment Institute, for his introduction of sandwich method.We appreciate the kind offer of transgenic aspen from former Prof. N. Morohoshi, Tokyo University of Agriculture and Technology.

6. References

Bevan, M.; Flavell, R.B. & Chilton, M.-D. (1983) A Chimeric Antibiotic Resistance Gene as a Selectable Marker for Plant Cell Transformation. *Nature*, Vol.304, (February 1983),pp. 184-187.

Fujii, Y.; Parvez, S.S.; Parvez, M.M.; Ohmae, Y. & Iida, O. (2003). Screening of 239 Medicinal Plant Species for Allelopathic Activity Using the Sandwich Method. *Weed Biology and Management*, Vol. 3, pp.233-241

Fujii, Y.; Shibuya, T.; Nakatani, K.; Itani, T.; Hiradate, S; Parvez, M.M. (2004). Assessment Method for Allelopathic Effect from Leaf Litter Leaches. *Weed Biology and Management*, Vol. 4, pp.19-23

Golisz, A.;Lata, B.;Gawronski, S.W.&Fujii, Y. (2007)Specific and Total Activities of the Allelochemicals Identified in Buckwheat. *Weed Biology and Management*,Vol.7, pp. 164-171

Haggman, H.; Frey, A. D.; Ryynanen, L.; Aronen,T.; Julukunen-Tiitto, R.; Tiimonen, H.; Pihakaski-Maunsbach, K.; Jokipii, S.; Chen, H. & Kallio, P. T. (2003). Expression of *Vitreoscilla* Haemoglobin in Hybrid aspen (*Populus tremula* x *tremuloides*), *Plant Biotechnology Journal*, Vol.1,(March 2003), pp. 287-300.

Hinchee, M.; Zhang, C; Chang, S.; Cunningham, M.;Hammond, W.;Nehra, N.; Person, L.(2011). Biotech Eucalyptus can Sustainably Address Society's Need for Wood: the example of Freeze Tolerant Eucalyptus in the Southeastern U.S., *IUFRO Tree Biotechnology Conference 2011 "From genomes to integration and delivery" Conference Proceedings Abstracts*, pp.438-439, Arraial D'Ajuda,Bahia,Brazil, June 26-July02, 2011

Japan Biosafety Clearing-House (2007). High Cellulose Rich White Poplar trg300-1 and 2 (*AaXEG2, Poplus alba* L.) . 07-46P-0003 and -0004, 2007-3-22

Kalaitzandonakes, N. (2004). Another Look at Biotech Regulation,*Regulation*, Vol.27,pp.44-50.

Kawaoka, A.; Kaothien, P.; Yoshida, K.; Endo, S., Yamada, K. & Ebinuma, H.(2000).Functional Analysis of Tobacco LIM Protein Ntlim1 Involved in Lignin Biosynthesis, *The Plant Journal*, Vol.22, (April 2000), pp.289-301.

Kawaoka, A.; Nanto, K.; Ishii, K. & Ebinuma, H. (2006). Reduction of Lignin Content by Suppression of the Lim Domain Transcription Factor in *Eucalyptus camaldulensis*, *Silvae Genetica*, Vol. 55, (December 2006), pp.269-277.

Kikuchi, A; Kawaoka, A,; Shimazaki, T.; Yu, X.; Ebinuma, H. & Watanabe, K.N. (2006). Trait Stability and Environmental Biosafety Assessments on Three Transgenic Eucalyptus Lines (*Eucalyptus camaldulensis* Dehnh, codA 12-5B, codA 12-5C, codA 20-C) conferring salt tolerance. *Breeding Research*, Vol.8, pp.17-26

Kikuchi, A.; Watanabe, K.; Tanaka, Y.; Kamada, H. (2008). Recent Progress on Environmental Biosafety Assessment of Geneticaly Modified Trees and Floricultural Plants in Japan. *Plant Biotechnology*, Vol.25, No.1, (March 2008), pp.9-15, ISSN 1342-4580

Kikuchi, A.; Yu, X.; Shimazaki T., Kawaoka, A.; Ebinuma, H.; Watanabe, K.N. (2009). Allelopathy Assessments for the Environmental Biosafety of the Salt-Tolerant Transgenic *Eucalyptus camaldulensis*, Genotypes codA 12-5B, codA 12-5C, and codA 20C.J. Wood Science, Vol.55, No.2, (April 2009), pp.149-153

Kole, C. & Hall, T.C.(2008). *Transgenic Forest Tree Species*.: Wiley-Blackwell, ISBN 978-1-4051-6924-0, Oxford, UK

Lu, M.-Z.; Hu, J.-J. (2011). A Brief Overview of Field Testing and Commercial Application of Transgenic Trees in China. *IUFRO Tree Biotechnology Conference 2011 "From genomes to integration and delivery"* Conference Proceedings Abstracts, pp.438-439, Arraial D'Ajuda,Bahia,Brazil, June 26-July02, 2011

Mclean, M. A. & Charrest, P. J.(2000). The Regulation of Transgenic Trees in North America, *Silvae Genetica*, Vol. 49, (December 2000), pp.233-239

Olsen, R. A.; Odham, G. & Linderberg, G.(1971). Aromatic Substances in Leaves of *Populus tremula* Inhibiters of Mycorrhizal Fungi. *Physiol. Plant.*, Vol.25,(March 1971), pp.122-129.

Redenbaugh, K. & McHughen, A.(2004). Regulatory Challenges Reduce Opportunities for Horticultural Biotechnology, *Calif. Agric.*,Vol. 58, pp.106-119.

Sekine, T.; Sugano, M.; Majid, A. & Fujii, Y. (2007). Antifungal Effects of Volatile Compunds from Black Zira (*Bunium persicum*) and Other Spices and Herbs. *J. Chem. Eco.* , Vol.33, pp.2123-2132.

Shimazaki, T.; Kikuchi, A.; Matsunaga, E.; Nanto, K.; Shimada, T. & Watanabe, K.N. (2009). Establishment of a Homogenized Method for Environmental Biosafety Assessments of Transgenic Plants. Plant Biotechnology, Vol.26, No.1, (March 2009), pp.143-148, ISSN 1342-4580

Shiomi, M.; Asakawa, Y.; Fukumoto, F.; Hamaya, E.; Hasebe, A.; Ichikawa, H.; Matsuda, I.; Muramatsu, T.; Okada, M.; Sato, M.; Ukai, Y.; Yokoyama, K.; Motoyoshi, F.; Ohashi, Y.; Ugaki, M. & Noguchi, K. (1992). Evaluation of the Impact of the Release of Transgenic Tomato Plants with TMV Resistance on the Environment. *Bull Natl Inst Agro-Environment Science Jpn.*, Vol.8, pp.1-51, ISSN 0911-9450

Singh, A.; Danda, R.S. & Ralham, P.K.(1993). Performance of Wheat Varieties under Poplar Plantation in Panjab,*Agroforestry Systems*, Vol.22,(February 1993),pp.83-86.

Strauss, S.H. (2003). Regulation of Biotechnology as though Gene Function Mattered. *BioScience*,Vol.53, pp.453-454

Strauss, S.H.; Tan, H.; Boerjan, W. & Sedjo, R. (2009). Strangled at Birth? Forest Biotech and the Convention on Biological Diversity, *Nature Biotechnology*, Vol.27, (June 2009), pp.519-527.

Walter, C.; Charity, J.; Donaldson, L.; Grace, L.; Mcdoonald, A., Moller, R. & Wagner, A. (2004). Genetic Modification of Conifer Forestry: State of the Art and Future Potential- A Case Study with *Pinus radiate*. In: Kumar S, Fladung M (eds) *Molecular Genetics and Breeding of Forest Trees*. Food Products Press, pp.215-242, NY, U.S.A.

Watanabe, K. N., Taeb, M & Okusu, H. (2004).Japanese Controversies over Transgenic Crop Regulation,*Science*, Vol.305, No.5691,(September 2004), pp1572

Yahong, L.; Tsuji, Y.; Nishikubo, N.; Kajita, S. & Morohoshi, N. (2001). Analysis of Transgenic Poplar in Which the Expression of Peroxidase Gene is Suppressed. In: Morohoshi N, Komamine A (eds) *Molecular Breeding of Woody Plants*. Elsevier, Amsterdam-Tokyo, pp.195-204

Yamaguchi, H. & Fujii, Y. (1994). Allelopathyc of Cultivated and Weed Azuki Beans –
     Assessment of Allelopathic Activity by Plant Box Method - . *J. Weed Science
     Technology*, Vol. 33, pp.138-139, ISSN:0372798X
Zeng, R. S., Mallik, A. U. & Luos. M. (2008). *Allelopathy in Sustainable Agriculture and
     Forestry*. Springer-Verlag, 2008, ISBN 0387773363.

Permissions

The contributors of this book come from diverse backgrounds, making this book a truly international effort. This book will bring forth new frontiers with its revolutionizing research information and detailed analysis of the nascent developments around the world.

We would like to thank Assoc. Prof. Yelda Özden Çiftçi, for lending her expertise to make the book truly unique. She has played a crucial role in the development of this book. Without her invaluable contribution this book wouldn't have been possible. She has made vital efforts to compile up to date information on the varied aspects of this subject to make this book a valuable addition to the collection of many professionals and students.

This book was conceptualized with the vision of imparting up-to-date information and advanced data in this field. To ensure the same, a matchless editorial board was set up. Every individual on the board went through rigorous rounds of assessment to prove their worth. After which they invested a large part of their time researching and compiling the most relevant data for our readers. Conferences and sessions were held from time to time between the editorial board and the contributing authors to present the data in the most comprehensible form. The editorial team has worked tirelessly to provide valuable and valid information to help people across the globe.

Every chapter published in this book has been scrutinized by our experts. Their significance has been extensively debated. The topics covered herein carry significant findings which will fuel the growth of the discipline. They may even be implemented as practical applications or may be referred to as a beginning point for another development. Chapters in this book were first published by InTech; hereby published with permission under the Creative Commons Attribution License or equivalent.

The editorial board has been involved in producing this book since its inception. They have spent rigorous hours researching and exploring the diverse topics which have resulted in the successful publishing of this book. They have passed on their knowledge of decades through this book. To expedite this challenging task, the publisher supported the team at every step. A small team of assistant editors was also appointed to further simplify the editing procedure and attain best results for the readers.

Our editorial team has been hand-picked from every corner of the world. Their multi-ethnicity adds dynamic inputs to the discussions which result in innovative outcomes. These outcomes are then further discussed with the researchers and contributors who give their valuable feedback and opinion regarding the same. The feedback is then collaborated with the researches and they are edited in a comprehensive manner to aid the understanding of the subject.

Apart from the editorial board, the designing team has also invested a significant amount of their time in understanding the subject and creating the most relevant covers. They scrutinized every image to scout for the most suitable representation of the subject and create an appropriate cover for the book.

The publishing team has been involved in this book since its early stages. They were actively engaged in every process, be it collecting the data, connecting with the contributors or procuring relevant information. The team has been an ardent support to the editorial, designing and production team. Their endless efforts to recruit the best for this project, has resulted in the accomplishment of this book. They are a veteran in the field of academics and their pool of knowledge is as vast as their experience in printing. Their expertise and guidance has proved useful at every step. Their uncompromising quality standards have made this book an exceptional effort. Their encouragement from time to time has been an inspiration for everyone.

The publisher and the editorial board hope that this book will prove to be a valuable piece of knowledge for researchers, students, practitioners and scholars across the globe.

List of Contributors

Yingying Meng, Hongyu Li, Tao Zhao, Chunyu Zhang and Bin Liu
Institute of Crop Science, Chinese Academy of Agricultural Sciences, Beijing, China

Chentao Lin
Department of Molecular, Cell & Developmental Biology, University of California, Los Angeles, USA

Inna Abdeeva, Sergey Bruskin and Eleonora Piruzian
NI Vavilov Institute of General Genetics RAS, Moscow, Russsia

Rustam Abdeev
Center for Theoretical Problems of Physico-Chemical Pharmacology RAS, Moscow, Russsia

Sílvia Coimbra and Luís Gustavo Pereira
University of Porto, Faculty of Sciences, Biology Department and BioFIG, Portugal

Iratxe Urreta and Sonia Castañón
Neiker-Tecnalia, Vitoria Gasteiz, Spain

Alain Tissier
Department of Cell and Metabolic Biology, Leibniz Institute of Plant Biochemistry, Weinberg, Germany

Ke-Gang Li and De-Yu Xie
North Carolina State University, Department of Plant Biology, Raleigh, NC, USA

Corey D. Broeckling
Proteomics and Metabolomics Facility, Colorado State University, Fort Collins, CO, USA

Yugang Song, Haiyan Shen and Hong Wang
Graduate University of the Chinese Academy of Sciences, Beijing, China

Yan Liu, Huahong Wang, Jianlin Chen, Benye Liu and Hechun Ye
Key Laboratory of Photosynthesis and Environmental Molecular Physiology, Institute of Botany, the Chinese Academy of Sciences, Beijing, China

Zhenqiu Li
Hebei University, Baoding, China

Juliano Lino Ferreira, Geraldo Magela de Almeida Cançado and Tesfahun Alemu Setotaw
Plant Biotechnology Laboratory, Agricultural Research Agency of Minas Gerais - EPAMIG, Caldas – MG, Brazil

Aluízio Borém and Wellington Silva Gomes
Department of Crop Science, Universidade Federal de Viçosa, Viçosa-MG, Brazil

Pusta Dana Liana
University of Agricultural Sciences and Veterinary Medicine Cluj-Napoca, Romania

Lilya Kopertekh and Joachim Schiemann
Julius Kuehn Institute, Federal Research Centre for Cultivated Plants (JKI), Institute for Biosafety of Genetically Modified Plants, Quedlinburg, Germany

Katsuaki Ishii and Toru Taniguchi
Forest Bio-Research Center, Forestry and Forest Products Research Institue, Japan

Akiyoshi Kawaoka
Nippon Paper Institute, Pulp and Paper Research Laboratory, Japan

Printed in the USA
CPSIA information can be obtained
at www.ICGtesting.com
JSHW011420221024
72173JS00004B/612